Innovative Waste Solutions

Cutting-Edge Techniques for a Greener World

Lovell Fuller

Innovative Waste Solutions

TABLE OF CONTENTS

Chapter 1: Revolutionary Waste Reduction Strategies

Source Reduction Techniques

The essence of source reduction lies in its proactive approach to waste management, focusing on the root of the issue rather than the aftermath. Unlike recycling or composting, which deal with waste after it's been created, source reduction aims to prevent waste from being generated in the first place. This approach not only reduces the burden on waste management systems but also conserves resources and energy, leading to a more sustainable and efficient use of materials.

One of the fundamental strategies for source reduction is the redesign of products and packaging to minimize waste. This can involve altering the materials used in production, which might include substituting non-recyclable materials with recyclable or biodegradable ones. It also involves designing products with durability and reusability in mind. For example, consider a shift from single-use packaging to multi-use containers that can be returned and refilled. This concept has seen successful implementation in various sectors, including the beverage industry, where companies have introduced glass bottles that are returned, sterilized, and reused multiple times before being recycled.

Another critical aspect of source reduction is conducting waste audits. These audits help identify the types of waste generated and their origins, providing valuable insights into potential areas for waste reduction. By understanding where waste originates within a system, organizations can make informed decisions to streamline processes and reduce excess. For instance, a manufacturing plant might discover that a significant portion of its waste stems from excess packaging materials. Armed with this knowledge, they can work with suppliers to reduce packaging or switch to more sustainable options.

Consumer behavior also plays a significant role in source reduction. Public awareness campaigns and education can drive changes in consumer habits, encouraging individuals to make conscious choices that minimize waste. Simple actions, such as opting for products with minimal packaging, choosing reusable over disposable items, and supporting companies with sustainable practices, can collectively have a substantial impact on waste reduction. For businesses, engaging consumers through incentives for returning or reusing products can foster a culture of sustainability. Some companies offer discounts or rewards for customers who bring their own containers, effectively reducing packaging waste.

Case studies of successful reduction programs provide practical examples of how organizations have implemented source reduction strategies. One notable example is a multinational electronics company that revamped its product design to reduce the number of components and simplify assembly. This not only lowered production costs but also resulted in less waste generated during manufacturing and at the end of the product's life cycle. Additionally, the company established a take-back program, encouraging consumers to return old devices for recycling and refurbishing, further minimizing electronic waste.

Another successful initiative is observed in the food industry, where a major retailer implemented a program to reduce food waste. By analyzing purchase and inventory data, the retailer optimized its supply chain to better match supply with demand, significantly reducing the amount of unsold food that went to waste. Furthermore, unsellable but edible food was donated to local food banks, turning potential waste into a valuable resource for the community.

Source reduction is not limited to large corporations; small businesses and individuals can also make significant contributions. A small café, for example, might switch to cloth napkins and reusable cutlery, reducing the reliance on disposable items. Homeowners can practice source reduction by composting kitchen scraps, choosing products with less packaging, and repairing rather than replacing household items. These actions, while seemingly small, contribute to a larger shift towards sustainability and resource conservation.

In the realm of legislation, governments can play a pivotal role in promoting source reduction through policies and regulations. Implementing standards for product design that prioritize recyclability and reusability can drive industry-wide change. Additionally, imposing taxes on single-use products and offering incentives for sustainable practices can encourage businesses and consumers to adopt source reduction strategies.

Collaborative efforts between governments, businesses, and consumers are essential for the widespread adoption of source reduction techniques. By fostering partnerships and sharing best practices, stakeholders can develop innovative solutions that address the unique challenges of waste management across different sectors and regions. For example, a city government might partner with local businesses to implement a shared logistics system that reduces packaging waste by consolidating deliveries and minimizing transportation emissions.

The benefits of source reduction extend beyond environmental impact. Businesses that adopt these practices often see financial gains through cost savings and increased efficiency. Consumers benefit from more sustainable products and services, and communities benefit from reduced waste, lower pollution levels, and a healthier environment. By prioritizing source reduction, society can move towards a more sustainable future where resources are used wisely, waste is minimized, and environmental impacts are significantly reduced.

Product Redesign for Minimal Waste

Redesigning products to minimize waste is a transformative approach that addresses sustainability at its core. It involves rethinking the entire lifecycle of a product, from its inception to its disposal, with a focus on reducing environmental impact. This process requires a shift in mindset, where companies prioritize not only profit but also ecological responsibility and resource conservation.

The journey begins with evaluating the materials used in production. Traditional materials, often chosen for cost efficiency or durability, may not always align with sustainable practices. For instance, plastics have been a staple in manufacturing due to their versatility and low cost, yet they contribute significantly to environmental pollution. By exploring alternative materials such as bioplastics, recycled content, or biodegradable composites, companies can significantly reduce the ecological footprint of their products. These materials not only offer a sustainable alternative but also open doors to innovation in product design and functionality.

Durability and longevity are key aspects of product redesign. A product designed to last longer inherently reduces waste, as it decreases the frequency of replacement. This approach challenges the prevailing culture of disposability and planned obsolescence, encouraging consumers to invest in quality over quantity. Some companies have embraced this philosophy by offering repair services or designing products with modular components that can be easily replaced or upgraded, extending the product's life and reducing waste.

Packaging is another critical area where redesign can have a substantial impact. Excessive packaging contributes to significant waste, much of which ends up in landfills. By minimizing packaging materials and opting for recyclable or compostable options, companies can address this issue head-on. Innovations such as edible packaging, plant-based materials, and minimalist design not only reduce waste but also enhance the consumer experience by aligning with growing environmental consciousness.

Functionality and multi-use design are also central to minimizing waste through product redesign. Products that serve multiple purposes or are easily adaptable to different uses reduce the need for additional items, thereby conserving resources. This concept is evident in the rise of multifunctional furniture, collapsible containers, and versatile clothing, which cater to consumers' desires for efficiency and practicality.

The principles of cradle-to-cradle design further enhance efforts to minimize waste. This concept focuses on creating products that can be fully reclaimed or re-used at the end of their lifecycle, essentially closing the loop on waste. By designing products with their eventual recycling or upcycling in mind, companies can ensure that materials are kept within the economic system rather than being discarded. This approach requires collaboration across the supply chain to establish systems for collecting, processing, and reintroducing materials into production.

Companies that embrace product redesign for minimal waste often see benefits beyond environmental impact. Consumers are increasingly drawn to brands that demonstrate a commitment to sustainability, valuing transparency and ethical practices. By communicating these values and integrating them into their brand identity, companies can differentiate themselves in a competitive market. This shift not only attracts environmentally conscious consumers but also fosters loyalty and trust.

The process of product redesign is not without its challenges. It requires an investment in research and development, as well as a willingness to experiment and take risks. Companies may face initial costs associated with sourcing sustainable materials or revamping production processes. However, these investments often pay off in the long term, as sustainable practices lead to cost savings, improved efficiency, and a stronger market position.

Collaboration and knowledge-sharing play a crucial role in advancing product redesign for minimal waste. Industry partnerships, cross-sector collaborations, and open-access platforms can facilitate the exchange of ideas and best practices. By working together, companies can leverage collective expertise to address complex challenges and drive innovation in sustainable design.

Consumers also have a role to play in supporting product redesign efforts. By making informed choices and prioritizing products that align with sustainable values, consumers can influence market trends and encourage companies to adopt more responsible practices. Public demand for transparency and accountability can drive companies to take meaningful action towards reducing waste.

Incorporating minimal waste principles into product design is a vital step towards a sustainable future. It requires a holistic approach that considers environmental impact at every stage of a product's lifecycle. By embracing innovation, collaboration, and consumer engagement, companies can create products that meet the needs of today while preserving resources for future generations. This transformative approach not only addresses the urgent challenges of waste management but also paves the way for a more sustainable and equitable world.

Waste Audits and Efficiency Improvements

Waste audits serve as a diagnostic tool in the quest for efficiency and sustainability. By systematically examining waste streams, businesses and organizations can identify inefficiencies, uncover opportunities for cost savings, and make informed decisions that align with environmental goals. This proactive approach not only reduces waste but also optimizes processes, ultimately leading to a more sustainable operation.

The process begins with a comprehensive assessment of waste generation within a facility. This involves gathering data on the types and quantities of waste produced, as well as pinpointing their sources. By involving various departments in this assessment, a holistic view of the waste management landscape emerges. Employees from different areas can provide insights into specific waste-generating activities, ensuring a thorough understanding of the entire system.

An essential component of a waste audit is the physical sorting and analysis of waste. This hands-on approach allows auditors to categorize waste into different streams, such as recyclables, compostables, and landfill-bound materials. By examining the contents of waste bins, auditors can uncover patterns and identify areas where waste reduction efforts should be concentrated. For instance, if a significant portion of waste consists of recyclable materials, it may indicate a need for improved recycling practices or employee education.

Once the data is collected, the next step is analyzing it to identify trends and opportunities for improvement. This analysis often reveals surprising insights, such as unexpected sources of waste or inefficiencies in existing processes. For example, a manufacturing facility might discover that a substantial amount of waste results from excess packaging materials. Armed with this information, the facility can work with suppliers to reduce packaging or switch to more sustainable alternatives.

Efficiency improvements often stem from the insights gained through waste audits. By targeting specific areas for intervention, organizations can implement targeted strategies to reduce waste and enhance operational efficiency. One common approach is optimizing inventory management to avoid overproduction and excess waste. By aligning production schedules with demand forecasts, companies can minimize waste resulting from unsold products.

Another area ripe for efficiency improvements is process optimization. Waste audits can identify bottlenecks or redundancies in workflows, enabling organizations to streamline operations and reduce waste generation. For instance, a restaurant may find that a significant portion of food waste comes from over-preparation. By adjusting portion sizes or refining menu offerings, the restaurant can reduce food waste while also improving cost efficiency.

Employee engagement is a critical factor in the success of waste audits and subsequent efficiency improvements. By involving employees in the audit process, organizations foster a culture of sustainability and accountability. Training sessions and workshops can educate staff on best practices for waste reduction, empowering them to contribute actively to sustainability efforts. Recognizing and rewarding employees for their contributions can further motivate them to embrace sustainable practices.

Collaboration with external partners and stakeholders can enhance the effectiveness of waste audits. Organizations can work with waste management companies, suppliers, and industry peers to share best practices and develop innovative solutions. For example, a business consortium might collaborate to create a shared logistics system that reduces transportation emissions and packaging waste. By leveraging collective expertise and resources, these partnerships can drive significant improvements in waste management.

Regulatory frameworks also play a role in shaping waste audit practices and efficiency improvements. Compliance with local and national regulations can serve as a catalyst for organizations to conduct waste audits and implement sustainable practices. In some cases, regulations may require organizations to report on their waste generation and reduction efforts, providing an added incentive to conduct thorough audits and pursue efficiency improvements.

The benefits of waste audits extend beyond environmental impact. Organizations that successfully reduce waste often see financial gains through cost savings and improved resource utilization. By identifying and eliminating inefficiencies, businesses can enhance their bottom line while also meeting sustainability goals. Additionally, companies that prioritize waste reduction can enhance their brand reputation, attracting customers and partners who value environmental responsibility.

Implementing a continuous improvement approach is crucial for maintaining the momentum of waste reduction efforts. Waste audits should not be a one-time exercise but rather an ongoing process of evaluation and refinement. By regularly assessing waste generation and revisiting efficiency strategies, organizations can adapt to changing circumstances and continue to make progress towards sustainability goals. This iterative approach ensures that waste reduction remains a priority and that organizations remain agile in the face of evolving challenges.

In conclusion, waste audits and efficiency improvements are powerful tools for organizations seeking to reduce waste and enhance sustainability. By systematically examining waste streams, identifying inefficiencies, and implementing targeted strategies, businesses can achieve significant environmental and financial benefits. Through collaboration, employee engagement, and a commitment to continuous improvement, organizations can create a culture of sustainability that drives long-term success and fosters a healthier planet for future generations.

Consumer Behavior and Waste Reduction

Understanding consumer behavior is pivotal in addressing waste reduction, as individuals' choices and habits significantly influence the amount of waste generated. By examining what drives consumer decisions and how these can be redirected towards more sustainable practices, we can foster a culture of responsibility and minimize waste at its origin. This approach not only involves educating consumers but also reshaping the environment in which they make decisions, creating a seamless transition towards more sustainable lifestyles.

Firstly, the role of consumer education cannot be overstated. Many individuals are unaware of the environmental impact of their choices, from the products they buy to the way they dispose of them. By raising awareness about the environmental footprint of different products and the benefits of sustainable alternatives, consumers can make informed decisions. Educational campaigns, workshops, and public information drives can be effective tools in this regard. For instance, a campaign highlighting the lifecycle of a plastic bottle—from production to disposal—can illustrate its environmental cost and promote reusable alternatives.

Alongside education, providing consumers with clear, accessible information at the point of purchase is essential. This can include labeling that indicates the environmental impact of a product or its packaging, as well as information on how to properly dispose of it. Eco-labels and certifications can guide consumers toward choosing products that align with their values, making it easier for them to contribute to waste reduction efforts.

Convenience plays a crucial role in shaping consumer behavior. For many, the ease of purchasing and disposing of products outweighs environmental considerations. To counteract this, businesses and policymakers can work together to make sustainable choices more convenient. This might involve developing infrastructure for reusable packaging, such as deposit-return schemes for bottles or refill stations for household products. By reducing the hassle associated with sustainable practices, more consumers are likely to adopt them.

Social influences also impact consumer behavior. People are often swayed by the actions and opinions of those around them, whether through direct interaction or the subtle cues of societal norms. Leveraging social dynamics can be a powerful tool in promoting waste reduction. Public figures, influencers, and community leaders can set examples by adopting and advocating for sustainable practices. Additionally, initiatives that create a sense of community around sustainability, such as local zero-waste challenges or community clean-up events, can foster collective action and reinforce positive behaviors.

Moreover, financial incentives can effectively motivate consumers to reduce waste. Discounts, rewards, or tax incentives for purchasing sustainable products or participating in recycling programs can encourage more environmentally friendly choices. For example, some grocery stores offer discounts to customers who bring their own bags, thus reducing plastic bag usage. Similarly, cities might provide tax breaks or rebates for households that achieve significant waste reduction.

However, consumer behavior is not solely driven by rational choices; emotions and perceptions play a significant role. Many consumers experience cognitive dissonance when their actions do not align with their environmental values, leading to discomfort and a desire for change. Businesses can tap into this by marketing products in a way that resonates emotionally with consumers, highlighting stories of sustainability and the positive impact of waste reduction.

Product design also influences consumer habits. Products designed for durability, repairability, and reusability encourage consumers to move away from disposable culture. By offering warranties, repair services, or trade-in programs, companies can extend the lifecycle of their products, reducing waste and building consumer loyalty. For example, a clothing retailer might offer repair workshops or recycling initiatives where customers can return old garments for a discount on future purchases.

Retail environments can be structured to promote sustainable choices. The placement of products, the design of store layouts, and the messaging used in advertising all contribute to consumer decision-making. Retailers can highlight sustainable products through prominent displays, special promotions, or dedicated sections in their stores. By making these products more visible and attractive, retailers can nudge consumers toward more sustainable purchases.

Finally, policy interventions play an indispensable role in shaping consumer behavior and waste reduction. Policymakers can enact regulations that encourage sustainable practices, such as banning single-use plastics, mandating recycling, or setting targets for waste reduction. These regulations can create a framework that supports and amplifies individual efforts, making sustainable choices the norm rather than the exception.

The interplay between consumer behavior and waste reduction is complex, involving a multitude of factors that influence decision-making. By understanding and addressing these factors, we can create a supportive environment that empowers consumers to make choices that align with their environmental values. Through education, incentives, and collaboration, we can drive a shift towards sustainability that begins at the individual level but resonates across society. This collective effort holds the promise of a future where waste is minimized, resources are conserved, and the planet is preserved for generations to come.

Case Studies of Successful Reduction Programs

Case studies of successful waste reduction programs provide tangible examples of how innovative strategies can lead to significant environmental and economic benefits. Each program, unique in its approach, offers valuable insights into the practical application of waste reduction techniques across various industries. By examining these real-world successes, we gain a deeper understanding of the potential and scalability of these initiatives.

One noteworthy case is the initiative undertaken by a global electronics manufacturer that set a precedent for sustainable production. Recognizing the massive waste generated by electronic devices, the company launched a comprehensive program focused on product redesign and lifecycle management. They invested in research and development to create modular devices, allowing consumers to easily replace or upgrade components instead of purchasing entirely new products. This approach not only reduced electronic waste but also extended the lifespan of their products, fostering customer loyalty and trust. The company complemented this with a robust take-back program, offering incentives for consumers to return old devices for refurbishment and recycling. As a result, they significantly reduced their waste output and set a benchmark for the industry.

In the food sector, a major retailer implemented a groundbreaking program to combat food waste, a pervasive issue with environmental and social implications. By leveraging advanced data analytics, the retailer was able to optimize inventory management, closely aligning supply with consumer demand. This minimized the occurrence of unsold perishables, drastically cutting down on waste. Additionally, they developed partnerships with local food banks and charities, ensuring that surplus food was redirected to those in need rather than being discarded. This initiative not only reduced waste but also addressed food insecurity, showcasing a model that has since been replicated by other retailers worldwide.

The fashion industry, notorious for its environmental impact, has also seen successful waste reduction programs. A leading fashion brand took a bold step by launching a circular fashion initiative, emphasizing the reuse and repurposing of garments. They introduced a buy-back scheme, encouraging customers to return unwanted clothing in exchange for store credit. These items were then either refurbished for resale or recycled into new materials. The brand also partnered with textile recycling firms to develop innovative recycling techniques that break down fibers for reuse. This closed-loop system not only reduced textile waste but also resonated with eco-conscious consumers, enhancing the brand's reputation and aligning with evolving market trends.

In the manufacturing realm, a prominent automotive company embarked on a zero-waste-to-landfill initiative. They conducted extensive waste audits to identify waste streams and areas for improvement within their production processes. Through a combination of source reduction, recycling, and innovative waste-to-energy solutions, the company achieved significant waste reduction milestones. Key to their success was the engagement of employees at all levels, fostering a culture of sustainability and continuous improvement. By transforming waste into a valuable resource, the company not only minimized its environmental footprint but also realized cost savings and operational efficiencies.

Municipalities have also played a crucial role in successful waste reduction efforts. A city in Europe implemented a comprehensive waste management strategy that integrated community engagement, policy changes, and technological innovations. They introduced a pay-as-you-throw program, incentivizing residents to reduce waste by charging based on the volume of trash disposed. Coupled with extensive recycling and composting facilities, the city achieved impressive waste diversion rates. Public education campaigns and community workshops further empowered residents to adopt sustainable practices, creating a collective sense of responsibility and pride in their city's achievements.

In the hospitality industry, a luxury hotel chain launched a sustainability program aimed at minimizing waste and conserving resources. They implemented a series of practices, such as eliminating single-use plastics, optimizing water and energy use, and sourcing local, sustainable products. An innovative aspect of their program was the introduction of a guest participation initiative, where guests could opt into environmental activities and earn rewards. This not only raised awareness but also engaged guests in the hotel's sustainability journey, enhancing their overall experience and satisfaction.

Educational institutions have also made significant strides in waste reduction. A university in North America established a comprehensive sustainability initiative that encompassed waste reduction, energy conservation, and sustainable food systems. They implemented campus-wide composting and recycling programs, reducing landfill waste by a substantial margin. The university also incorporated sustainability into the curriculum, providing students with hands-on opportunities to engage in waste reduction projects. This holistic approach not only achieved impressive waste reduction results but also fostered a culture of sustainability among students, staff, and the wider community.

These case studies highlight the diverse approaches and strategies that organizations across different sectors have employed to successfully reduce waste. Common themes among these initiatives include innovation, collaboration, and an unwavering commitment to sustainability. By sharing best practices and lessons learned, these programs inspire others to pursue similar paths, demonstrating the tangible benefits of waste reduction efforts. As more organizations adopt these strategies, we move closer to a more sustainable future, where waste is minimized, resources are conserved, and the environmental impact is significantly reduced.

Chapter 2: Advanced Recycling Technologies

Mechanical vs. Chemical Recycling

Recycling is an essential component of sustainable waste management, offering a means to conserve resources and reduce environmental impact. Within this realm, mechanical and chemical recycling stand as two distinct approaches, each with its unique processes, benefits, and challenges. Understanding the differences between these methods and their applications can illuminate pathways to more effective recycling strategies.

Mechanical recycling is the more traditional and widely practiced form, involving the physical processing of waste materials to produce new products. This process typically includes collection, sorting, cleaning, shredding, and melting, before the material is finally reformed into new items. For example, plastic bottles are often collected, cleaned to remove contaminants, and then shredded into flakes. These flakes are melted and remolded into new plastic products. Mechanical recycling is particularly effective for certain types of plastics, metals, and glass, where the material can be processed multiple times without significant degradation of quality.

One of the primary advantages of mechanical recycling is its relatively low cost and energy efficiency compared to chemical recycling. The process is less complex and can be implemented at a larger scale with existing infrastructure. It also has a lower carbon footprint, as it does not require the chemical reactions and energy-intensive processes associated with chemical recycling. However, mechanical recycling is not without its limitations. The quality of recycled materials can diminish with each cycle, particularly for plastics, which can become brittle or discolored. Contaminants such as food residue or mixed materials can also pose challenges, reducing the efficiency and effectiveness of the process.

Chemical recycling, on the other hand, involves breaking down waste materials into their basic chemical components, which can then be reconstituted into new materials. This process often employs chemical reactions or solvents to depolymerize plastics, converting them back into monomers or other chemical building blocks. These can then be used to produce new, virgin-quality materials. Chemical recycling offers the potential to handle a broader range of plastics, including those that are difficult to recycle mechanically, such as multi-layered packaging or heavily contaminated materials.

The primary advantage of chemical recycling lies in its ability to produce high-quality, virgin-equivalent materials. This can significantly extend the lifecycle of plastics, reducing the need for fossil fuel-derived virgin materials. Additionally, chemical recycling can potentially handle a wider variety of plastic waste streams, contributing to higher overall recycling rates. However, the process is often more energy-intensive and costly than mechanical recycling, requiring advanced technology and infrastructure. It also presents challenges in terms of scalability and economic viability, as current chemical recycling facilities are limited in capacity and geographic distribution.

The decision between mechanical and chemical recycling depends on several factors, including the type of material, the level of contamination, and the intended application of the recycled product. In many cases, a combination of both methods may be the most effective approach, leveraging the strengths of each to maximize recycling outcomes. For example, mechanical recycling can be used for materials that are relatively clean and straightforward to process, while chemical recycling can address more complex waste streams that would otherwise be destined for landfill or incineration.

In practice, successful recycling programs often involve a comprehensive and integrated approach that encompasses both mechanical and chemical methods. This can include investments in technology and infrastructure, as well as collaboration with industry partners and stakeholders to develop efficient and scalable solutions. Policy and regulation also play a critical role in shaping the recycling landscape, with governments implementing standards and incentives to encourage the adoption of advanced recycling technologies.

Education and public awareness are equally important components of effective recycling strategies. Consumers play a vital role in the recycling process, and their participation is crucial for success. By educating the public about the differences between mechanical and chemical recycling, as well as the importance of proper sorting and disposal, we can enhance recycling rates and reduce contamination. Community engagement and outreach programs can further support these efforts, fostering a culture of sustainability and responsibility.

Innovation and research continue to drive advancements in recycling technology, with ongoing efforts to improve the efficiency, cost-effectiveness, and environmental impact of both mechanical and chemical processes. New developments in enzyme-based recycling, solvent extraction, and pyrolysis, for instance, hold promise for expanding the capabilities of chemical recycling, while innovations in sorting technology and material separation enhance the effectiveness of mechanical recycling.

The intersection of mechanical and chemical recycling represents a dynamic and evolving field, with the potential to significantly impact waste management and resource conservation. By understanding the strengths and limitations of each approach, we can develop more comprehensive and sustainable recycling systems that meet the challenges of today and tomorrow. Through collaboration, innovation, and a commitment to continuous improvement, we can transform waste into valuable resources, contributing to a circular economy and a healthier planet.

Innovations in Plastic Recycling

The world is grappling with the ever-growing challenge of plastic waste, a problem that has spurred a wave of innovations in recycling technologies. These advancements are not only transforming how we manage plastic waste but are also paving the way for a more sustainable future. By exploring cutting-edge methodologies and technologies, we can redefine the lifecycle of plastics and mitigate their environmental impact.

One of the most promising innovations in plastic recycling is the development of advanced sorting technologies. Traditional recycling facilities often struggle with the challenge of sorting different types of plastics, which vary in their chemical compositions and properties. Innovations like near-infrared (NIR) spectroscopy and artificial vision systems are revolutionizing this process. NIR technology can rapidly and accurately identify and categorize plastics based on their spectral signatures, allowing for more precise sorting. Similarly, artificial vision systems equipped with machine learning algorithms can recognize and sort plastics with remarkable efficiency, reducing contamination and increasing the purity of recycled materials.

Another groundbreaking advancement is the emergence of enzyme-based recycling. Unlike conventional methods that rely on mechanical or chemical processes, enzyme-based recycling employs biological catalysts to break down plastics at the molecular level. Researchers have discovered specific enzymes that can depolymerize plastics like polyethylene terephthalate (PET) into their basic monomers. These monomers can then be repurposed to create new, high-quality plastic products. The enzyme-based approach offers several advantages, including reduced energy consumption, lower carbon emissions, and the ability to handle a broader range of plastic types, including those that are traditionally difficult to recycle.

Pyrolysis, a thermal decomposition process, is also gaining traction as an innovative solution for plastic recycling. Pyrolysis involves heating plastics in the absence of oxygen to break them down into smaller molecules, which can then be converted into fuels or raw materials for new plastic production. This process has the potential to transform mixed and contaminated plastic waste into valuable resources, providing a viable alternative to traditional disposal methods. Some companies are already implementing pyrolysis on an industrial scale, turning plastic waste into synthetic crude oil, which can be refined into various petrochemical products.

The concept of chemical recycling is being reimagined through innovations such as solvent-based depolymerization. This method utilizes solvents to selectively dissolve and recover polymers from plastic waste, leaving behind impurities and additives. The recovered polymers can be precipitated and purified for reuse, offering a pathway to recycle plastics that are otherwise challenging to process mechanically. Solvent-based depolymerization holds promise for dealing with complex waste streams, such as multi-layered packaging and composite materials, which are typically not recyclable through traditional methods.

Recycling initiatives are also focusing on the development of bioplastics, which are derived from renewable resources such as plant starches or microbial fermentation. While bioplastics are designed to be biodegradable or compostable, their integration into existing recycling systems can be complex. Innovations in recycling bioplastics involve designing processes that can effectively separate and process these materials alongside conventional plastics, ensuring that they do not contaminate recycling streams. By improving the compatibility of bioplastics with current recycling infrastructure, we can enhance their viability as a sustainable alternative to traditional plastics.

Community-driven recycling innovations are emerging as well, particularly in regions with limited access to formal recycling infrastructure. Localized solutions, such as small-scale plastic melting machines and community-based recycling hubs, empower individuals and communities to take control of plastic waste management. These initiatives often involve collaboration with local governments, NGOs, and businesses to create sustainable models that can be replicated in other areas. By fostering a sense of ownership and responsibility, community-driven innovations contribute to a broader cultural shift towards sustainability.

The role of digital technology in plastic recycling cannot be overlooked. Digital platforms and apps are being developed to connect consumers with recycling services, providing real-time information on recycling locations, guidelines, and incentives. These platforms can also facilitate the exchange of recycled materials, creating a marketplace for recycled plastics that enhances their value and accessibility. By leveraging digital technology, we can streamline recycling processes and engage a wider audience in the effort to reduce plastic waste.

While these innovations hold great promise, challenges remain in scaling and integrating them into existing systems. Issues such as economic feasibility, regulatory barriers, and market acceptance must be addressed to realize the full potential of these technologies. Collaborative efforts between governments, industry stakeholders, and research institutions are crucial to overcoming these hurdles and driving the widespread adoption of innovative recycling solutions.

Ultimately, the future of plastic recycling lies in a combination of technological advancements, systemic changes, and cultural shifts. By embracing innovation and fostering collaboration, we can create a more sustainable and resilient recycling ecosystem that not only addresses the current plastic waste crisis but also anticipates future challenges. As we continue to push the boundaries of what is possible, the innovations in plastic recycling offer a beacon of hope for a cleaner, more sustainable world.

E-Waste Recycling: Techniques and Challenges

E-waste, or electronic waste, represents one of the fastest-growing waste streams globally, driven by rapid technological advancements and consumer demand for the latest gadgets. The recycling of e-waste is crucial, not only to recover valuable materials but also to mitigate the environmental and health hazards posed by improper disposal. Despite its importance, e-waste recycling presents unique challenges, ranging from the complex composition of electronic products to the economic and logistical hurdles of processing them efficiently.

The first step in effective e-waste recycling is the collection and sorting of discarded electronic devices. This process is complicated by the sheer diversity of products, each containing a mix of metals, plastics, glass, and other materials. Collection systems need to be efficient and accessible to ensure that e-waste is diverted from landfills and directed toward recycling facilities. Community collection events, designated drop-off points, and take-back programs offered by manufacturers are some of the strategies employed to increase collection rates. However, the success of these initiatives often hinges on public awareness and participation, underscoring the need for educational campaigns to inform consumers about the importance of e-waste recycling.

Once collected, e-waste undergoes a series of processes to extract valuable materials. Manual dismantling is often the initial step, where devices are carefully taken apart to separate different components. Skilled workers remove batteries, circuit boards, and other parts that require specialized handling. This labor-intensive process is crucial for minimizing contamination and maximizing material recovery, but it also poses challenges in terms of labor costs and safety. Workers need proper training and protective equipment to handle hazardous substances, such as lead, mercury, and cadmium, commonly found in electronics.

Shredding and sorting follow dismantling, where materials are broken down into smaller pieces and separated by type. Advanced technologies, such as magnetic and eddy current separation, are used to isolate ferrous and non-ferrous metals, while optical sorting systems help distinguish between different types of plastics. These processes are essential for ensuring that the materials can be effectively recycled and reused. However, the complexity of e-waste, with its intricate mix of materials, often results in losses and inefficiencies, highlighting the need for continuous innovation in sorting technologies.

Chemical processes, such as hydrometallurgical and pyrometallurgical methods, are employed to recover precious metals from e-waste. Hydrometallurgy involves using aqueous solutions to dissolve metals, which are then precipitated and purified. This method is particularly effective for extracting gold, silver, and copper from circuit boards and connectors. Pyrometallurgy, on the other hand, uses high temperatures to smelt metals, facilitating the recovery of materials like aluminum and zinc. While these processes are essential for recovering valuable resources, they can be energy-intensive and environmentally taxing, necessitating stringent controls and advancements to reduce their impact.

The economic viability of e-waste recycling is another significant challenge. The fluctuating prices of recovered materials and the high costs associated with collection and processing can make it difficult for recycling operations to be profitable. To address this, some countries have implemented extended producer responsibility (EPR) programs, which mandate that manufacturers finance the collection and recycling of their products. These programs aim to internalize the environmental costs of e-waste, incentivizing manufacturers to design more sustainable and easily recyclable products.

Informal recycling sectors pose additional challenges, particularly in developing countries where e-waste is often processed in unsafe and environmentally damaging ways. Informal recyclers may use rudimentary techniques, such as open burning and acid leaching, to extract metals, exposing themselves and the environment to toxic substances. Addressing this issue requires a multifaceted approach, including strengthening regulations, providing training and support to informal recyclers, and creating incentives for formal recycling operations. International cooperation is also vital to prevent the illegal shipment of e-waste to countries with lax environmental standards.

Innovation and research are key to overcoming the challenges of e-waste recycling. Emerging technologies, such as bioleaching, offer promising alternatives to traditional methods. Bioleaching uses microorganisms to extract metals from e-waste, presenting a potentially more environmentally friendly and cost-effective solution. Researchers are also exploring the development of modular electronics, which would be easier to disassemble and recycle, thereby reducing the complexity and cost of e-waste processing.

Public policy and consumer behavior play critical roles in the success of e-waste recycling. Governments can enact regulations and provide incentives to promote responsible recycling practices, while consumers can contribute by making informed choices and participating in recycling programs. Education and awareness campaigns are essential for fostering a culture of sustainability, encouraging individuals and organizations to prioritize the recycling and responsible disposal of electronic devices.

Despite the challenges, e-waste recycling presents significant opportunities for resource recovery, environmental protection, and economic development. By advancing recycling technologies, strengthening regulatory frameworks, and fostering collaboration across sectors, we can create a more sustainable approach to managing electronic waste. As the demand for electronics continues to grow, the importance of effective e-waste recycling will only increase, making it an essential component of a circular economy and a sustainable future.

The Role of Artificial Intelligence in Recycling

Artificial intelligence (AI) is revolutionizing the recycling industry, introducing efficiencies and innovations that were previously unimaginable. By integrating AI technology into recycling processes, we can significantly enhance the accuracy and speed of sorting materials, optimize waste management systems, and ultimately increase recycling rates. Understanding the role of AI in recycling necessitates a deep dive into its applications, benefits, and potential challenges.

One of the most transformative applications of AI in recycling is in the realm of automated sorting systems. These systems utilize machine learning algorithms and computer vision to identify and categorize materials with remarkable precision. Traditional recycling facilities often rely on manual sorting, which is labor-intensive and prone to human error. In contrast, AI-powered systems can process vast amounts of waste with high speed and accuracy. For instance, AI-equipped robotic arms can distinguish between different types of plastics, metals, and paper, sorting them into the appropriate streams. This not only reduces contamination in recycling batches but also improves the quality of the materials recovered.

The implementation of AI in sorting facilities is complemented by the use of sensors and data analytics. Sensors embedded in waste management systems collect data on the composition and flow of materials. This information is then analyzed using AI algorithms to optimize sorting processes and identify inefficiencies. By providing real-time insights, AI enables facility operators to make informed decisions, adjust operations, and maximize resource recovery. The ability to adapt and improve sorting processes dynamically is a significant advantage over traditional methods, which often lack flexibility and responsiveness.

Beyond sorting, AI is playing a pivotal role in predictive maintenance and operational efficiency within recycling facilities. By analyzing data from machinery and equipment, AI systems can predict when maintenance is needed, preventing costly breakdowns and minimizing downtime. This proactive approach not only extends the lifespan of equipment but also ensures that recycling operations run smoothly and efficiently. Furthermore, AI can optimize energy consumption within facilities, reducing the environmental footprint of recycling processes and contributing to sustainability goals.

AI's impact on recycling extends to supply chain management and logistics. Advanced algorithms can forecast waste generation patterns, enabling more efficient scheduling of collection and transportation. By predicting when and where waste will accumulate, recycling companies can optimize their routes, reduce fuel consumption, and lower operational costs. This level of precision is especially valuable in urban areas, where waste generation is high, and efficient logistics are essential. AI-driven logistics systems also enhance the coordination between collection, sorting, and processing facilities, streamlining the entire recycling chain.

Consumer engagement is another area where AI is making significant strides. AI-powered apps and platforms are being developed to educate and incentivize individuals to participate in recycling programs. These digital tools can provide personalized recommendations, track recycling habits, and offer rewards for sustainable behavior. By leveraging AI to engage consumers, recycling programs can increase public participation and foster a culture of environmental responsibility. Additionally, AI can analyze consumer data to identify trends and preferences, informing the design of more effective recycling campaigns and initiatives.

Despite the numerous benefits, the integration of AI into recycling presents challenges that must be addressed. One primary concern is the initial cost of implementing AI technology, which can be prohibitive for smaller facilities and developing regions. Overcoming this barrier requires investment and support from governments and industry stakeholders to ensure equitable access to AI-driven solutions. Additionally, the reliance on data and algorithms raises questions about privacy and data security. Recycling facilities must implement robust measures to protect sensitive information and maintain public trust.

The ethical implications of AI in recycling also warrant consideration. As AI systems replace manual labor in sorting and processing, there is a potential impact on employment within the industry. It is crucial to approach this transition with a focus on reskilling and upskilling workers, providing opportunities for them to adapt to new roles within AI-enhanced facilities. By prioritizing human capital alongside technological advancement, the recycling industry can navigate this shift responsibly and sustainably.

Looking ahead, the potential for AI in recycling is vast and largely untapped. Continued research and development are essential to unlocking new applications and improving existing technologies. Collaboration between technology providers, recycling companies, and policymakers will be key to driving innovation and ensuring that AI's benefits are realized across the entire recycling ecosystem.

AI's role in recycling is poised to reshape the industry, offering unprecedented opportunities for efficiency, sustainability, and innovation. By embracing AI and addressing its challenges, we can create a more effective and resilient recycling system that meets the demands of a rapidly changing world. As AI technology continues to evolve, its integration into recycling processes will undoubtedly play a central role in achieving a circular economy and a more sustainable future.

Global Examples of Cutting-Edge Recycling Facilities

Recycling facilities around the world are at the forefront of innovation, continually advancing techniques to manage waste more effectively. These cutting-edge facilities serve as benchmarks for the industry, showcasing what is possible with the right combination of technology, policy, and community engagement. By examining global examples, we can glean insights into the future of recycling and the potential for transformative environmental impact.

In Sweden, the city of Eskilstuna has become a model of sustainable waste management with its innovative recycling park, ReTuna. This facility combines a traditional recycling center with a shopping mall dedicated entirely to repaired and upcycled goods. As residents drop off their unwanted items, skilled workers sort through the materials, identifying those that can be refurbished or repurposed. Craftspeople and entrepreneurs then transform these items into sellable products, which are displayed and sold in various shops within the mall. ReTuna not only diverts waste from landfills but also promotes a circular economy by giving new life to discarded items. This facility has gained international attention, inspiring similar initiatives in other countries.

Japan, a nation known for its commitment to recycling, boasts the Kamikatsu Zero Waste Center, which exemplifies community involvement in waste management. Kamikatsu, a small town on Shikoku Island, has implemented a rigorous waste sorting system that requires residents to separate their waste into 45 different categories. This meticulous approach allows for highly efficient recycling and composting, significantly reducing landfill contributions. The Zero Waste Center serves as both a recycling facility and an educational hub, hosting workshops and tours to raise awareness about sustainable practices. Kamikatsu's success is largely due to the active participation of its residents, demonstrating the power of community engagement in achieving ambitious environmental goals.

In the Netherlands, the city of Amsterdam is home to AEB Amsterdam, one of the world's most advanced waste-to-energy facilities. AEB Amsterdam processes approximately 1.4 million tons of waste annually, converting it into electricity and heat for the city. The facility employs state-of-the-art technology, including high-efficiency incinerators and advanced flue gas cleaning systems, to minimize emissions and maximize energy recovery. The facility also recovers valuable materials, such as metals and minerals, from the incineration process, further contributing to resource conservation. AEB Amsterdam's integrated approach to waste management exemplifies how cities can harness energy from waste while minimizing environmental impact.

In Singapore, the Semakau Landfill stands out as an innovative solution to the challenges of land scarcity and waste management. Unlike traditional landfills, Semakau is an offshore facility, created by enclosing a part of the sea with a rock bund. Waste is incinerated on the mainland, and the ash is transported to Semakau, where it is used to fill the enclosed area. The landfill is designed to minimize environmental impact, with measures in place to protect marine life and water quality. Semakau also serves as a recreational area, with lush greenery and wildlife habitats that attract visitors. This dual-purpose approach highlights the potential for creative solutions in waste management, particularly in regions with limited land availability.

In South Korea, the Sudokwon Landfill Site Management Corporation (SLC) operates a facility that integrates waste management with resource recovery and renewable energy production. The facility processes waste from the Seoul metropolitan area, capturing landfill gas to generate electricity.

SLC also operates a leachate treatment plant and a composting facility, ensuring that all aspects of waste management are addressed. The site is home to a research and development center that focuses on advancing waste management technologies and practices. Sudokwon's comprehensive approach demonstrates the potential for landfill sites to be transformed into centers of innovation and sustainability.

In Canada, the Edmonton Waste Management Centre is a beacon of innovation, featuring one of North America's most advanced waste processing facilities. The center includes a co-composting plant, a construction and demolition waste recycling facility, and a materials recovery facility, among others. A standout feature is the Enerkem Alberta Biofuels facility, which converts non-recyclable waste into biofuels and chemicals through a gasification process. This integration of waste-to-biofuel technology reduces landfill contributions and provides a renewable energy source. The Edmonton Waste Management Centre exemplifies the potential for collaboration between municipalities, private companies, and research institutions to drive advancements in recycling and resource recovery.

In Australia, the Northern Adelaide Waste Management Authority (NAWMA) operates a state-of-the-art recycling facility that incorporates cutting-edge sorting technology. The facility uses optical sorters, ballistic separators, and magnets to efficiently process mixed recyclables, achieving high purity rates. NAWMA's commitment to innovation extends to its education and outreach programs, which engage the community in recycling efforts and promote sustainable practices. By combining advanced technology with public engagement, NAWMA sets a high standard for recycling facilities in the region.

These global examples illustrate the diverse approaches to cutting-edge recycling, each tailored to the specific needs and challenges of their respective regions. Common themes emerge, such as the integration of technology, community involvement, and a commitment to sustainability. By learning from these examples, other regions can develop and implement strategies that address their unique waste management challenges, ultimately contributing to a more sustainable future. As recycling facilities continue to evolve, they will play a critical role in the transition to a circular economy, where waste is minimized, resources are conserved, and environmental impact is reduced.

Chapter 3: Composting: Traditional Methods with Modern Twists

Community and Industrial Composting Innovations

Composting has long been recognized as a natural and effective method for recycling organic waste, transforming it into nutrient-rich soil amendments that enhance soil health and fertility. As communities and industries alike seek sustainable waste management solutions, innovations in composting have emerged, offering new techniques and systems that maximize efficiency and environmental benefits.

In urban areas, where space is limited and waste generation is high, community composting initiatives have gained traction as a grassroots approach to waste diversion. These projects often involve collaboration between local governments, non-profit organizations, and residents to create shared composting facilities. By pooling resources and knowledge, communities can establish composting hubs that accommodate food scraps, yard waste, and other organic materials. These hubs not only reduce landfill contributions but also foster a sense of community and environmental stewardship among participants. Some cities have even introduced curbside composting programs, where organic waste is collected alongside recyclables and trash, streamlining the process for residents and increasing participation rates.

One innovative approach to community composting is the use of bokashi, a fermentation process that breaks down organic waste through the action of beneficial microorganisms. Unlike traditional composting, which requires aeration and turning, bokashi involves the anaerobic fermentation of waste in sealed containers. This method is particularly well-suited for urban environments, as it produces minimal odors and can process a wide range of organic materials, including meat and dairy. The fermented waste can then be buried in the soil or added to a traditional composting system for further decomposition. By offering an efficient and space-saving solution, bokashi has become a popular choice for apartment dwellers and community gardens.

Industries, too, are embracing composting innovations to manage organic waste generated by production processes. One notable example is the use of in-vessel composting systems, which provide a controlled environment for the rapid decomposition of organic materials. These systems consist of enclosed containers or drums that maintain optimal conditions for microbial activity, such as temperature, moisture, and aeration. In-vessel composting is highly efficient, capable of processing large volumes of organic waste in a matter of weeks, and is particularly beneficial for industries with consistent waste streams, such as food processing plants and agricultural operations.

Another industrial innovation is the adoption of vermicomposting, a technique that utilizes worms to break down organic matter into nutrient-rich castings. Vermicomposting systems can be scaled to accommodate various waste volumes, making them suitable for both small businesses and large industrial operations. The process is highly efficient, producing high-quality compost with minimal labor and energy inputs. Additionally, the worms themselves can be harvested as a source of protein for animal feed, creating a closed-loop system that maximizes resource recovery.

The integration of technology into composting processes has also spurred significant advancements. Automated monitoring systems equipped with sensors and data analytics are now being used to optimize composting conditions, ensuring that temperature, moisture, and oxygen levels are maintained at ideal levels. These systems can provide real-time feedback and alerts, allowing operators to make adjustments as needed to accelerate decomposition and improve compost quality. By enhancing process control and efficiency, technology-driven composting solutions are helping to meet the growing demand for sustainable waste management in both community and industrial settings.

Despite the numerous benefits, the expansion of composting initiatives faces challenges, particularly in terms of infrastructure and public participation. Establishing and maintaining composting facilities requires investment and collaboration among stakeholders, including local governments, businesses, and community groups. Educating the public about the importance of composting and proper waste separation is also crucial to ensuring the success of these programs. Addressing these challenges involves not only technical solutions but also social and policy measures that incentivize and support composting efforts at all levels.

To overcome barriers and drive innovation, many regions are exploring partnerships and pilot projects that demonstrate the viability and benefits of composting. For example, some municipalities have partnered with local farms to process urban organic waste, creating a symbiotic relationship that supports local agriculture and reduces waste disposal costs. Others have launched composting pilot programs in schools and public institutions, using them as educational tools to raise awareness and inspire future generations to adopt sustainable practices.

Policy and regulation play a critical role in promoting composting innovations. By implementing mandates and incentives for organic waste diversion, governments can encourage businesses and communities to adopt composting solutions. Policies that support the development of composting infrastructure, such as grants and tax incentives, can also help overcome financial barriers and accelerate the adoption of new technologies and systems.

As composting innovations continue to evolve, they offer promising solutions to the pressing challenges of waste management and environmental sustainability. By embracing these advancements and fostering collaboration across sectors, we can unlock the full potential of composting as a tool for resource recovery, soil health, and climate resilience. Whether through community-driven initiatives or industrial-scale systems, composting has the power to transform waste into valuable resources, contributing to a more sustainable and regenerative future.

Vermicomposting and Other Biological Methods

Vermicomposting, a method that relies on the diligent work of worms to transform organic waste into nutrient-rich compost, stands at the forefront of biological waste recycling techniques. This process not only addresses the challenge of organic waste disposal but also enriches soil health, making it a valuable tool for gardeners, farmers, and environmentalists alike. Alongside vermicomposting, a host of other biological methods are gaining traction, each leveraging the power of nature to recycle waste efficiently and sustainably.

The essence of vermicomposting lies in its simplicity and efficiency. By introducing specific species of worms, most commonly Eisenia fetida, or red wigglers, into a controlled environment filled with organic matter, one can witness the transformation of waste into rich, dark compost. These worms consume organic waste, digest it, and excrete castings, which are laden with beneficial microbes and nutrients. The result is a natural fertilizer that enhances soil structure, aeration, and water retention, providing a sustainable alternative to chemical fertilizers.

Setting up a vermicomposting system is accessible and requires minimal investment. It begins with selecting or constructing a suitable bin, which can range from a simple plastic container to a more elaborate wooden structure, depending on the scale of composting desired. The bin should provide adequate ventilation to prevent odors and maintain appropriate moisture levels. Bedding material, such as shredded newspaper, cardboard, or coconut coir, is essential to create a comfortable habitat for the worms. Once the bin is prepared, adding the worms and gradually introducing organic waste, such as fruit and vegetable scraps, coffee grounds, and eggshells, initiates the vermicomposting process.

The care and maintenance of a vermicomposting system revolve around monitoring conditions to ensure the worms thrive. This includes maintaining a consistent moisture level, akin to a damp sponge, and avoiding extreme temperatures that could harm the worms. Regularly checking the pH level is also crucial, as overly acidic or alkaline conditions can be detrimental. Adding crushed eggshells or calcium carbonate can help neutralize acidity. Additionally, avoiding certain waste materials, such as meat, dairy, and oily foods, is essential to prevent odors and attract pests. With proper care, a vermicomposting system can produce high-quality compost in a matter of months.

Vermicomposting offers numerous benefits beyond waste reduction and soil enhancement. It is an educational tool, providing insights into natural processes and fostering environmental awareness. Schools and community groups often use vermicomposting to engage students and residents in sustainable practices, demonstrating the lifecycle of organic matter and the importance of waste management. Furthermore, vermicomposting is scalable, making it suitable for small households, urban gardens, and large agricultural operations alike. Its versatility and adaptability have made it a popular choice for those seeking to minimize their environmental footprint.

In addition to vermicomposting, other biological methods are gaining attention for their effectiveness in recycling organic waste. One such method is aerobic composting, which relies on microorganisms to decompose organic matter in the presence of oxygen. This process generates heat, accelerating the breakdown of waste and producing finished compost more quickly than traditional methods. Aerobic composting is often used in large-scale operations, such as municipal composting facilities, where windrows or aerated static piles allow for efficient processing of significant volumes of waste. The resulting compost is used in landscaping, agriculture, and land restoration projects, contributing to soil health and sustainability.

Anaerobic digestion represents another biological approach, utilizing microorganisms to break down organic matter in the absence of oxygen. This process occurs in sealed tanks or digesters, where waste is converted into biogas and digestate. The biogas, composed primarily of methane and carbon dioxide, can be captured and used as a renewable energy source, powering homes, businesses, and vehicles. The digestate, a nutrient-rich byproduct, can be applied to fields as a natural fertilizer. Anaerobic digestion is particularly valuable for managing wet and high-moisture organic waste, such as food scraps and agricultural residues, offering a dual benefit of waste reduction and energy production.

Composting with fungi is an emerging method that harnesses the natural decomposing abilities of fungi to break down complex organic materials. Certain fungi, such as oyster mushrooms, can degrade lignin and cellulose, components of woody materials and agricultural waste, making them valuable allies in composting. Inoculating compost piles with fungal spores can accelerate decomposition and enhance the nutrient profile of the finished compost. Fungal composting is being explored in various applications, from agricultural waste management to bioremediation, where fungi are used to break down pollutants and restore contaminated soils.

The integration of biological methods into waste management strategies presents opportunities for innovation and sustainability. Combining techniques, such as vermicomposting and anaerobic digestion, can optimize resource recovery and minimize waste. For instance, pre-treating organic waste with worms before anaerobic digestion can enhance biogas production and improve the efficiency of the overall process.

These synergies highlight the potential for creative solutions that leverage the strengths of different biological methods.

As the demand for sustainable waste management solutions grows, the role of biological methods in recycling will continue to expand. Their ability to transform waste into valuable resources aligns with the principles of a circular economy, where materials are continuously cycled and reused. By embracing these natural processes, individuals, communities, and industries can contribute to environmental conservation and resilience.

The journey into biological waste recycling is one of discovery and innovation, where the power of nature is harnessed to address modern challenges. Whether through vermicomposting, aerobic composting, anaerobic digestion, or fungal composting, these methods offer pathways to a more sustainable and regenerative future. Through continued exploration and adoption, we can unlock the full potential of biological recycling and nurture a world where waste is no longer a burden but a resource.

Technological Advances in Aerobic and Anaerobic Composting

Composting has evolved from a simple, natural process into a sophisticated technique, thanks to technological advancements in both aerobic and anaerobic methods. These innovations have significantly enhanced the efficiency and scalability of composting, allowing it to accommodate diverse waste streams and meet increasing demands for sustainable waste management.

Aerobic composting, which relies on the presence of oxygen to decompose organic matter, has seen remarkable progress through the integration of technology. One of the most significant advancements is the use of aerated static piles. This method involves the forced aeration of compost piles using a system of pipes and blowers, which supply oxygen directly into the mass. By controlling airflow, these systems maintain optimal conditions for microbial activity, accelerating decomposition and reducing the time required to produce mature compost. The result is a more efficient process that minimizes labor and can handle larger volumes of waste, making it particularly suitable for municipal and agricultural applications.

To further enhance aerobic composting, temperature monitoring and control systems have been developed. These systems use sensors to track the internal temperature of compost piles, ensuring that it remains within the ideal range for microbial activity. By maintaining temperatures between 55°C and 65°C, pathogens are effectively destroyed, and the process is expedited. Automated systems can adjust aeration rates based on real-time data, optimizing conditions and improving the consistency and quality of the finished product.

Windrow composting, another aerobic technique, has also benefited from technological advancements. This method involves arranging waste in long rows, or windrows, which are periodically turned to introduce oxygen and regulate temperature. The development of specialized equipment, such as windrow turners, has streamlined this process. These machines efficiently mix and aerate the compost, reducing manual labor and ensuring uniform decomposition. The ability to process vast quantities of organic waste quickly makes windrow composting a preferred option for large-scale operations.

On the other hand, anaerobic composting, which takes place in the absence of oxygen, has been revolutionized by the adoption of anaerobic digestion technology. This process involves the breakdown of organic material by microorganisms in a sealed environment, producing biogas and digestate. Anaerobic digesters have been enhanced with sophisticated monitoring and control systems that optimize conditions for microbial activity. These systems regulate factors such as temperature, pH, and nutrient balance, maximizing biogas production and improving process efficiency.

The biogas generated through anaerobic digestion is a valuable renewable energy source, composed primarily of methane and carbon dioxide. Technological advancements have improved the capture and utilization of biogas, allowing it to be used for electricity generation, heating, and even as a vehicle fuel. Upgrading biogas to biomethane, a purified form of methane, has become increasingly feasible, providing a sustainable alternative to fossil fuels. The digestate, a nutrient-rich byproduct, can be used as a natural fertilizer, completing the cycle of resource recovery.

Integration of co-digestion techniques has further expanded the potential of anaerobic composting. Co-digestion involves the simultaneous processing of multiple types of organic waste, such as food scraps, agricultural residues, and wastewater sludge, within the same digester. This approach enhances the efficiency of anaerobic digestion by balancing nutrient levels and increasing biogas yield. Technological innovations have enabled precise control of co-digestion processes, optimizing the synergy between different waste streams and maximizing resource recovery.

The advent of hybrid systems that combine aerobic and anaerobic processes represents a significant leap forward in composting technology. These systems harness the strengths of both methods to achieve superior results. For instance, pre-treating organic waste with aerobic composting can reduce its volume and improve its suitability for anaerobic digestion. The subsequent anaerobic digestion process then generates biogas and further stabilizes the material. Hybrid systems offer flexibility and efficiency, making them ideal for complex waste streams and large-scale operations seeking to minimize environmental impact.

Data analytics and digital platforms are playing an increasingly important role in advancing composting technology. By collecting and analyzing data from composting operations, these platforms provide valuable insights into process performance and efficiency. Operators can use this information to make informed decisions, optimize resource allocation, and address issues proactively. Digital platforms also facilitate knowledge sharing and collaboration, enabling stakeholders to learn from each other and adopt best practices.

Despite the remarkable progress, challenges remain in the widespread adoption of advanced composting technologies. The initial cost of implementing these systems can be high, particularly for small-scale operations and developing regions. Overcoming this barrier requires investment and support from governments and industry stakeholders, as well as innovative financing models that make technology accessible to all. Additionally, public education and awareness are crucial to promoting the benefits of composting and encouraging participation in organic waste diversion programs.

Policy and regulation also play a crucial role in fostering technological advancements in composting. Governments can incentivize the adoption of advanced systems through grants, tax incentives, and regulatory frameworks that prioritize waste reduction and resource recovery. By creating a supportive environment for innovation, policymakers can drive progress and ensure that composting technologies continue to evolve and meet the changing needs of society.

As we look to the future, the potential for technological advances in composting is vast and largely untapped. Continued research and development will unlock new applications and improve existing systems, enhancing the role of composting in sustainable waste management. By embracing these innovations and fostering collaboration across sectors, we can create a more efficient and resilient waste management system that contributes to environmental sustainability and resource conservation. Through technological progress, composting can transform from a traditional practice into a cornerstone of the circular economy, where waste is minimized, resources are conserved, and the health of our planet is prioritized.

The Impact of Composting on Soil Health

Composting, a process as old as agriculture itself, has become a modern cornerstone for enhancing soil health and promoting sustainable agriculture. By returning nutrients to the earth through the decomposition of organic matter, composting plays a pivotal role in maintaining and improving soil quality. The benefits of composting extend far beyond waste reduction, offering a natural solution to many of the challenges faced by farmers and gardeners alike.

Healthy soil is the foundation of any thriving ecosystem. It supports plant growth, regulates water, and sustains biodiversity. Yet, soil degradation remains a pressing global issue, driven by factors such as intensive farming, deforestation, and climate change. Composting provides a means to combat these challenges by enriching the soil with organic matter, improving its structure, fertility, and resilience.

One of the primary benefits of composting is its ability to enhance soil structure. Compost adds organic matter to the soil, which increases its porosity and helps retain moisture. This improved structure allows for better root penetration and enhances the soil's capacity to hold water, reducing the need for irrigation. The organic matter in compost also fosters the development of soil aggregates, which are clumps of soil particles bound together by organic substances. These aggregates improve soil stability and reduce erosion, protecting the land from the impacts of wind and water.

Compost is a rich source of nutrients essential for plant growth, including nitrogen, phosphorus, and potassium. Unlike synthetic fertilizers, which provide nutrients in concentrated forms, compost releases nutrients slowly as it decomposes. This slow release aligns with the natural nutrient uptake patterns of plants, reducing the risk of nutrient leaching and runoff that can pollute waterways. Additionally, compost contains trace elements and micronutrients that are often lacking in chemical fertilizers, contributing to balanced plant nutrition and soil fertility.

The presence of beneficial microorganisms in compost is another critical factor in improving soil health. These microorganisms, including bacteria, fungi, and actinomycetes, play a vital role in breaking down organic matter and cycling nutrients within the soil. As compost is added to the soil, it introduces and stimulates microbial activity, enhancing nutrient availability and promoting plant growth. The microbial diversity in compost can also suppress soil-borne diseases by outcompeting or inhibiting pathogenic organisms, reducing the need for chemical pesticides.

Composting also contributes to carbon sequestration, a process that captures atmospheric carbon dioxide and stores it in the soil. By incorporating organic matter into the soil, composting enhances its ability to act as a carbon sink, mitigating the impacts of climate change. This carbon storage not only benefits the environment but also improves soil health by increasing organic carbon content, which is crucial for sustaining soil fertility and productivity.

In addition to these benefits, composting helps to improve soil pH, creating a more favorable environment for plant growth. Many soils suffer from acidity or alkalinity, which can limit the availability of nutrients and hinder plant development. Compost acts as a buffer, helping to neutralize soil pH and create conditions conducive to healthy plant growth. This buffering capacity is particularly valuable in areas with naturally acidic or alkaline soils, where composting can significantly enhance agricultural productivity.

The use of compost in soil management practices can also reduce the reliance on synthetic inputs, such as fertilizers and pesticides. By improving soil health and fertility naturally, composting supports sustainable agriculture and reduces the environmental impact of farming. This reduction in chemical inputs not only benefits the environment but also lowers production costs for farmers, promoting economic sustainability.

Despite the clear advantages, there are still challenges to the widespread adoption of composting as a soil health strategy. These challenges include the availability of organic materials, the time required for composting, and the need for education and technical support. Addressing these challenges requires a concerted effort from governments, industry stakeholders, and communities to promote composting and provide the necessary resources and infrastructure.

Innovative approaches to composting are helping to overcome some of these barriers. For instance, co-composting, which involves the simultaneous composting of multiple types of organic waste, can enhance the efficiency and effectiveness of the process. By combining materials with complementary properties, such as food waste and yard trimmings, co-composting can optimize nutrient balance and accelerate decomposition. Additionally, the use of technology in composting, such as automated monitoring systems and data analytics, is improving process control and efficiency, making composting more accessible and scalable.

Community composting initiatives are also playing a vital role in promoting composting and its benefits. By engaging residents in waste diversion and composting activities, these initiatives foster environmental awareness and stewardship. Community composting sites provide opportunities for education and skill-building, empowering individuals to contribute to soil health and sustainability. These grassroots efforts are essential for building a culture of composting and expanding its impact on soil health.

The journey to enhancing soil health through composting is one of collaboration and innovation. By harnessing the power of nature and integrating it with modern technology and practices, we can unlock the full potential of composting as a tool for sustainable agriculture and environmental conservation. As we continue to explore and adopt composting strategies, we pave the way for a more resilient and productive future, where soil health is preserved and celebrated as the foundation of life on Earth.

Incentives and Policies Supporting Composting

Governments and organizations worldwide increasingly recognize the critical role composting plays in sustainable waste management and environmental conservation. To accelerate the adoption of composting initiatives, various incentives and policies have been developed, aiming to support individuals, communities, and industries in their composting efforts. These measures not only encourage the reduction of organic waste sent to landfills but also promote the use of compost as a valuable resource for agriculture and landscaping.

One of the most effective ways governments have encouraged composting is through financial incentives. Tax credits and grants are offered to businesses and municipalities that invest in composting infrastructure and technology. These financial aids can help offset the costs of setting up composting facilities, purchasing equipment, and conducting training programs. For instance, some regions provide subsidies for purchasing compost bins for households or commercial composting systems, making the initial investment more feasible for small businesses and community groups.

Another approach involves implementing mandatory organic waste diversion programs. These policies require households and businesses to separate organic waste from general trash, ensuring that compostable materials are directed to appropriate facilities. Such mandates can significantly increase the volume of organic waste collected for composting, as seen in several cities that have successfully implemented curbside composting programs. These initiatives often include educational components to teach residents and businesses about the benefits of composting and how to participate effectively.

Landfill tipping fees, which are charges levied on the disposal of waste at landfills, also serve as an economic incentive for composting. By increasing the cost of landfill disposal, governments can motivate waste generators to seek alternative waste management strategies, such as composting. This approach not only reduces the financial burden on municipalities for landfill maintenance but also conserves landfill space, prolonging its lifespan.

Incentives for using compost in agriculture and landscaping further support composting efforts. Some regions offer subsidies or tax deductions to farmers and landscapers who purchase and apply compost to their land. These measures encourage the adoption of compost as a soil amendment, enhancing soil health and reducing reliance on chemical fertilizers. Additionally, governments may establish quality standards for compost products, ensuring that they meet specific criteria for use in agriculture and horticulture. By certifying compost quality, these standards build consumer confidence and stimulate market demand.

Policies that integrate composting into broader sustainability and waste management strategies are also essential. For example, zero waste initiatives, which aim to eliminate waste through reduction, reuse, and recycling, often include composting as a key component. By setting ambitious waste diversion targets, these initiatives drive the development of composting infrastructure and programs. Collaboration between government agencies, businesses, and community organizations is crucial to the success of such strategies, fostering a unified approach to waste reduction and resource recovery.

Public-private partnerships have emerged as a powerful tool for advancing composting initiatives. These collaborations leverage the strengths and resources of both sectors to develop and implement composting projects. For instance, a city government might partner with a private waste management company to establish a large-scale composting facility, sharing the costs and benefits. Such partnerships can enhance efficiency, innovation, and scalability, making composting accessible to a wider range of participants.

Education and outreach programs are vital components of policies supporting composting. By raising awareness about the environmental benefits of composting and providing practical guidance, these programs empower individuals and businesses to participate actively. Workshops, seminars, and informational campaigns can demystify the composting process, address common misconceptions, and highlight success stories. Engaging schools, community groups, and local leaders in these efforts can amplify their reach and impact, creating a culture of composting and environmental stewardship.

Regulatory frameworks play a crucial role in facilitating composting activities. Clear guidelines on composting practices, facility operation, and product standards ensure that composting is conducted safely and effectively. Regulations may cover aspects such as site selection, odor management, and pathogen control, protecting public health and the environment. By establishing a transparent and supportive regulatory environment, governments can encourage compliance and innovation in the composting sector.

Research and development initiatives are essential for advancing composting technologies and practices. Governments and research institutions can support studies that explore new composting methods, improve process efficiency, and assess the environmental impacts of composting. By investing in research, policymakers can identify best practices and develop evidence-based policies that drive the composting industry forward.

Finally, international cooperation and knowledge sharing are critical for scaling up composting efforts globally. By learning from successful programs and policies in other regions, governments can adapt and implement effective strategies in their own contexts. International organizations and networks provide platforms for exchanging information, experiences, and resources, promoting collaborative solutions to common challenges.

The combined impact of incentives and policies supporting composting is transformative, fostering a shift towards more sustainable waste management practices. By aligning economic, regulatory, and educational measures, governments and organizations can create an enabling environment for composting to thrive. As composting becomes more integrated into waste management systems, its benefits for the environment, economy, and society become increasingly apparent. Through continued commitment and collaboration, we can harness the potential of composting to build a more sustainable and resilient future for all.

Chapter 4: Waste-to-Energy: Transforming Waste into Power

The Science of Waste-to-Energy Conversion

Turning waste into energy is a concept that has evolved from a mere possibility to a critical component of modern waste management and energy production strategies. Waste-to-energy (WTE) conversion not only provides a solution to the growing problem of waste disposal but also contributes to the energy demands of our society. By harnessing the latent energy in waste materials, WTE technologies offer a sustainable approach to both energy generation and waste reduction, making them a key player in the quest for a circular economy.

The primary science behind waste-to-energy conversion involves the transformation of waste materials into usable forms of energy, typically heat, electricity, or fuel. This process is achieved through various methods, each leveraging different principles of physics and chemistry. Among the most prevalent technologies are incineration, gasification, pyrolysis, and anaerobic digestion, each offering unique advantages and applications.

Incineration is one of the oldest and most widely implemented waste-to-energy technologies. It involves the combustion of organic materials at high temperatures, converting waste into ash, flue gas, and heat. The heat generated is used to produce steam, which drives turbines to generate electricity. Modern incineration plants are equipped with advanced pollution control systems to minimize emissions and capture harmful substances. The efficiency of incineration, coupled with its ability to significantly reduce waste volume, makes it a popular choice for municipal solid waste management.

Gasification is a process that converts organic or fossil-based materials into syngas, a mixture of hydrogen, carbon monoxide, and some carbon dioxide. This is achieved by reacting the material at high temperatures with a controlled amount of oxygen or steam. Syngas can be used for electricity generation or as a chemical feedstock, offering a versatile and cleaner alternative to traditional fossil fuels. Gasification allows for greater control over the combustion process, resulting in lower emissions compared to direct incineration.

Pyrolysis is similar to gasification but occurs in the absence of oxygen. This thermal decomposition process breaks down organic materials into char, bio-oil, and syngas. The bio-oil can be refined into fuels or chemicals, while the syngas can be used for energy generation. Pyrolysis offers the advantage of producing valuable by-products that can be further processed, enhancing the economic viability of waste-to-energy conversion.

Anaerobic digestion is a biological process that involves the breakdown of organic matter by microorganisms in the absence of oxygen. This process produces biogas, primarily composed of methane and carbon dioxide, which can be used as a renewable energy source. The remaining digestate is a nutrient-rich material that can be used as a fertilizer. Anaerobic digestion is particularly suited for wet organic waste, such as food scraps and agricultural residues, and is often used in conjunction with other waste management strategies.

The integration of waste-to-energy technologies into waste management systems offers numerous environmental and economic benefits. By reducing the volume of waste sent to landfills, WTE technologies help mitigate the environmental impact of waste disposal, including the release of greenhouse gases and leachate. Furthermore, the energy produced from waste can offset the use of fossil fuels, contributing to energy security and reducing carbon emissions.

However, the implementation of waste-to-energy technologies is not without challenges. Public perception and acceptance can be significant hurdles, particularly concerning concerns about air pollution and health impacts. Ensuring that WTE facilities comply with stringent environmental regulations and employ the latest pollution control technologies is crucial for addressing these concerns and gaining public trust.

Economic factors also play a critical role in the adoption of waste-to-energy technologies. The initial investment for building WTE facilities can be substantial, and the feasibility of projects often depends on factors such as waste availability, energy market conditions, and government incentives. Developing comprehensive policies and support mechanisms that encourage investment and innovation in WTE technologies is essential for realizing their full potential.

Research and development continue to drive advancements in waste-to-energy technologies, improving efficiency, reducing costs, and expanding the range of applicable waste materials. Innovations such as plasma arc gasification and microbial fuel cells are emerging, offering promising alternatives for energy recovery from waste. These technologies leverage cutting-edge science to enhance the sustainability and effectiveness of waste-to-energy systems.

Moreover, the integration of waste-to-energy with other renewable energy sources, such as solar and wind, presents opportunities for creating hybrid energy systems. These systems can provide a more stable and resilient energy supply by balancing intermittent renewable sources with the continuous output of WTE facilities. Such integration aligns with the goals of a circular economy, where waste is considered a resource and energy production is diversified and sustainable.

The role of education and public engagement in promoting waste-to-energy technologies cannot be overstated. By raising awareness of the benefits and safety of WTE, communities can be encouraged to support and participate in these initiatives. Educational programs, stakeholder consultations, and transparent communication are vital for building trust and fostering collaboration between the public, industry, and governments.

International cooperation and knowledge exchange are also crucial for advancing waste-to-energy technologies globally. By sharing best practices, experiences, and technical expertise, countries can accelerate the adoption and optimization of WTE systems. Collaborative efforts can drive innovation, reduce costs, and address common challenges, ultimately contributing to global sustainability goals.

In conclusion, the science of waste-to-energy conversion represents a transformative approach to waste management and energy production. By embracing and advancing these technologies, we can turn the challenge of waste into an opportunity for sustainable development. As we continue to innovate and integrate WTE into our energy systems, we move closer to a future where waste is minimized, resources are conserved, and energy is produced in harmony with the environment. Through continued commitment and cooperation, waste-to-energy technologies can play a pivotal role in building a sustainable and resilient world.

Innovations in Incineration and Gasification

Incineration and gasification have long been central to waste management practices, offering solutions for reducing landfill dependency and generating energy. Recent innovations in both processes are transforming how we handle waste, enhancing efficiency, reducing emissions, and expanding the range of materials that can be processed. These advancements are crucial as the world grapples with mounting waste challenges and seeks more sustainable energy sources.

Incineration, the process of burning waste at high temperatures to reduce its volume and recover energy, has undergone significant improvements. Traditional incineration faced criticisms mainly due to emissions of pollutants such as dioxins and furans. However, modern incinerators are equipped with advanced flue gas cleaning systems that drastically reduce these harmful emissions. Technologies such as fabric filters, scrubbers, and catalytic converters are now standard, capturing particulate matter and neutralizing acidic gases. These innovations have made incineration a cleaner and more acceptable option in waste management.

The efficiency of energy recovery has also seen progress. Combined Heat and Power (CHP) systems are increasingly integrated into incineration plants, allowing the simultaneous production of electricity and heat. This dual output maximizes energy utilization from waste, making incineration more economically viable. The heat generated can be used in district heating systems, providing warmth to residential and industrial buildings, thereby enhancing the overall energy efficiency.

Moreover, the development of fluidized bed incineration represents a leap forward. This technology involves suspending waste particles in a hot air stream, which enhances combustion efficiency and allows for the processing of a wider variety of waste types, including those with high moisture content. The improved thermal efficiency and reduced emissions associated with fluidized bed systems make them particularly attractive for modern waste management facilities.

Gasification, a process that converts organic or fossil-based materials into syngas through partial oxidation, has seen equally transformative advancements. Unlike incineration, gasification operates in a low-oxygen environment, producing syngas that can be used as a cleaner alternative to fossil fuels. Recent innovations have focused on improving the quality and yield of syngas, making the process more efficient and economically attractive.

One significant development is the integration of plasma arc technology in gasification. By using extremely high temperatures generated by an electric arc, plasma gasification breaks down waste materials into their basic molecular components. This results in a cleaner syngas with fewer impurities, expanding its potential applications. Plasma gasification also produces a vitrified slag, a non-toxic byproduct that can be used in construction, thus addressing waste disposal issues.

Advancements in reactor design have further optimized gasification processes. The introduction of advanced feed systems and the use of catalysts have improved the conversion efficiency and reduced the energy input required. These innovations allow for the processing of a broader range of waste materials, including those with lower calorific values, making gasification a versatile solution for waste management.

The coupling of gasification with renewable energy sources is another promising innovation. By using solar or wind energy to provide the necessary heat for gasification, the carbon footprint of the process can be significantly reduced. This hybrid approach not only enhances sustainability but also stabilizes energy supply by providing a continuous output that complements intermittent renewable sources.

In addition to technological advancements, the integration of digital technology and data analytics is playing an increasingly important role in optimizing both incineration and gasification processes. Real-time monitoring systems and predictive maintenance algorithms improve operational efficiency and minimize downtime. By analyzing data from the combustion or gasification process, operators can make informed decisions to enhance performance and reduce emissions.

Public awareness and acceptance are crucial for the successful implementation of incineration and gasification technologies. Addressing concerns about emissions and health impacts requires transparent communication and strict adherence to environmental regulations. Educational initiatives that highlight the benefits and safety measures associated with these technologies can foster public trust and support.

The economic feasibility of incineration and gasification is influenced by factors such as waste availability, energy market conditions, and regulatory frameworks. Governments can play a pivotal role by providing incentives for the adoption of these technologies and by establishing policies that prioritize waste-to-energy solutions. By creating a supportive environment, policymakers can drive investment and innovation in the sector.

International collaboration and knowledge exchange are vital for advancing incineration and gasification technologies globally. Sharing best practices and technical expertise can help countries overcome common challenges and accelerate the adoption of these technologies. Collaborative efforts can also drive down costs and enhance the accessibility of advanced waste management solutions.

In conclusion, the innovations in incineration and gasification are reshaping the landscape of waste management and energy production. By embracing these advancements, we can address the dual challenges of waste disposal and energy demand in a sustainable manner. Through continued research, development, and collaboration, incineration and gasification can contribute significantly to a circular economy where waste is transformed into a valuable resource. The journey towards sustainable waste management and energy production is ongoing, with incineration and gasification at the forefront of this transformation. As we continue to innovate, these technologies will play an increasingly important role in building a sustainable and resilient future.

Biomass and Biogas Production Techniques

Biomass and biogas production techniques have emerged as key players in the quest for renewable energy, providing sustainable alternatives to fossil fuels while addressing waste management challenges. These methods leverage organic materials to produce energy, offering a dual benefit of resource recovery and environmental conservation. By understanding the science and practical applications of biomass and biogas technologies, we can harness their potential to contribute to a cleaner and more sustainable energy future.

Biomass refers to organic material that comes from plants and animals, which can be used as a renewable energy source. It includes wood, agricultural crops, and waste from forests, yards, or farms. The conversion of biomass into energy can be achieved through various processes, each with its own set of advantages and considerations.

One of the most common techniques for biomass conversion is direct combustion, where biomass is burned to produce heat. This heat can be used directly for industrial processes or to generate electricity by producing steam that drives turbines. Modern biomass combustion systems are designed to maximize efficiency and minimize emissions, incorporating technologies such as advanced boilers and emission control devices. The use of biomass for heating and electricity generation is particularly popular in regions with abundant agricultural and forestry residues, providing a cost-effective and sustainable energy solution.

Another method of biomass conversion is through biochemical processes, such as fermentation and anaerobic digestion. Fermentation involves the conversion of biomass into bioethanol, a renewable liquid fuel, through the action of microorganisms. This process is commonly used with sugar-rich crops like corn and sugarcane, where the sugars are fermented into alcohol. Bioethanol can be used as a direct fuel or blended with gasoline to power vehicles, offering a cleaner-burning alternative to conventional fuels.

Anaerobic digestion is a biological process that breaks down organic matter in the absence of oxygen, producing biogas and digestate. Biogas, primarily composed of methane and carbon dioxide, is a versatile energy source that can be used for electricity generation, heating, or as a vehicle fuel. The digestate, a nutrient-rich byproduct, can be used as a fertilizer, closing the nutrient loop and enhancing soil health. Anaerobic digestion is particularly suited for wet organic waste, such as food scraps and animal manure, making it an ideal solution for farms and municipalities seeking to manage organic waste sustainably.

Thermochemical processes also play a significant role in biomass conversion. Pyrolysis, for example, involves the thermal decomposition of biomass in the absence of oxygen, producing bio-oil, syngas, and char. Bio-oil can be refined into transportation fuels, chemicals, or used for heating, while syngas can be utilized for electricity generation or as a chemical feedstock. Char, a solid byproduct, can be used as a soil amendment or for carbon sequestration. The flexibility and diversity of products derived from pyrolysis make it an attractive option for biomass conversion.

Gasification is another thermochemical process that converts biomass into syngas through partial oxidation. This high-temperature process offers a cleaner and more efficient way to produce energy from biomass compared to direct combustion. The syngas produced can be used in combined cycle power plants to generate electricity with higher efficiency or processed into liquid fuels and chemicals. Gasification is particularly advantageous for large-scale energy production, providing a reliable and continuous energy supply.

The production of biogas, a key component of biomass energy, is gaining traction as a renewable energy solution. Biogas plants, which utilize anaerobic digestion, offer a sustainable way to manage organic waste while producing energy. These facilities can be found on farms, where animal manure is digested to produce biogas, or in municipal settings, treating food waste and sewage sludge. The versatility of biogas as an energy source, along with its ability to reduce methane emissions from waste decomposition, underscores its importance in the renewable energy landscape.

Recent innovations in biomass and biogas production techniques are enhancing their efficiency and scalability. The development of advanced enzymes and microorganisms is improving the conversion rates of biochemical processes, reducing the time and resources required. Technological advancements in reactor design and process optimization are also enabling more efficient and cost-effective biomass conversion. Additionally, integrated biorefineries, which combine multiple biomass conversion processes, are emerging as hubs for renewable energy and chemical production, maximizing resource utilization and economic viability.

The environmental benefits of biomass and biogas production are significant. By replacing fossil fuels with renewable biomass energy, we can reduce greenhouse gas emissions and mitigate climate change. Biomass energy systems also contribute to waste reduction and resource recovery, turning agricultural residues, forestry byproducts, and organic waste into valuable energy. These processes support a circular economy, where waste is minimized, and resources are reused, enhancing sustainability and resilience.

Economic considerations are crucial for the widespread adoption of biomass and biogas technologies. The cost of feedstock, technology, and infrastructure can influence the feasibility of biomass projects. Government incentives, such as subsidies, tax credits, and renewable energy mandates, play a vital role in promoting investment and development in the sector. By creating a supportive policy environment, governments can drive the growth of biomass and biogas industries, contributing to energy security and economic development.

Public awareness and education are essential for fostering acceptance and support for biomass and biogas technologies. By highlighting the benefits and addressing misconceptions, stakeholders can build trust and encourage participation in renewable energy initiatives. Community engagement and collaboration with industry and government can create a shared vision for a sustainable energy future, where biomass and biogas play a central role.

International cooperation and knowledge sharing are critical for advancing biomass and biogas production techniques globally. By exchanging best practices, research findings, and technical expertise, countries can accelerate the adoption and optimization of these technologies. Collaborative efforts can also address common challenges, such as feedstock availability and process efficiency, driving innovation and progress in the renewable energy sector.

In conclusion, biomass and biogas production techniques offer promising solutions for renewable energy generation and waste management. By leveraging organic materials and innovative technologies, we can transform waste into valuable energy resources, contributing to a sustainable and resilient energy future. As we continue to explore and develop these techniques, they will play an increasingly important role in addressing global energy challenges and building a cleaner, more sustainable world. Through continued research, development, and collaboration, biomass and biogas can become integral components of our energy systems, supporting a transition to a low-carbon economy.

Environmental Considerations and Regulatory Frameworks

Navigating the balance between industrial progress and environmental sustainability has become a pressing concern in today's world. As waste management and energy production continue to evolve, understanding the environmental considerations and regulatory frameworks that govern these fields is paramount. These frameworks are designed to mitigate the environmental impact while promoting sustainable practices, ensuring that development does not come at the cost of ecological integrity.

Environmental considerations in waste management and energy production encompass a broad range of factors. At the core is the need to minimize emissions and pollutants that can harm ecosystems and public health. This includes the release of greenhouse gases such as carbon dioxide and methane, which contribute to global warming, as well as pollutants like nitrogen oxides and particulate matter that can degrade air quality. Strategies to reduce these emissions are critical, as they not only protect the environment but also enhance the efficiency and sustainability of waste-to-energy processes.

One of the primary approaches to addressing these concerns is the implementation of advanced emission control technologies. In incineration plants, for example, systems such as electrostatic precipitators, scrubbers, and selective catalytic reduction are employed to capture and neutralize harmful substances before they are released into the atmosphere. These technologies are essential for ensuring that waste-to-energy facilities operate within the stringent limits set by environmental regulations.

Water usage and contamination are additional environmental considerations. Waste management processes, particularly those involving thermal treatments, can require significant amounts of water for cooling and cleaning. Ensuring that this water is used efficiently and treated appropriately before being discharged is crucial to prevent water pollution. Closed-loop water systems and advanced filtration technologies are increasingly being adopted to minimize water consumption and contamination, aligning waste management practices with broader sustainability goals.

The management of byproducts and residues is another critical aspect. Processes such as incineration and gasification produce ash and slag, which must be handled and disposed of responsibly to prevent soil and water contamination. Developing strategies for the safe disposal or beneficial reuse of these byproducts, such as in construction materials, is key to reducing the environmental footprint of waste-to-energy technologies.

Regulatory frameworks play a vital role in guiding and enforcing environmental protections. These regulations establish the standards and requirements that waste management and energy production facilities must adhere to, ensuring that they operate in an environmentally responsible manner. Regulatory bodies, both national and international, set guidelines for emissions, waste handling, and resource usage, providing a legal backbone for sustainable practice.

In many regions, regulations are shaped by international agreements and conventions aimed at protecting the environment. The Paris Agreement, for example, represents a global commitment to reducing greenhouse gas emissions and mitigating climate change. Such international frameworks influence national policies and encourage countries to adopt practices that align with global sustainability goals.

Compliance with environmental regulations requires a proactive approach, involving continuous monitoring and reporting. Facilities must implement robust monitoring systems to track emissions and resource usage, ensuring that they remain within permitted limits. Regular reporting to regulatory authorities not only demonstrates compliance but also fosters transparency and accountability. This process is crucial for building public trust and maintaining the social license to operate.

Incentives and support mechanisms are often embedded within regulatory frameworks to encourage the adoption of environmentally friendly technologies and practices. These can include tax credits, grants, and subsidies for facilities that invest in advanced emission control systems or renewable energy sources. By providing financial support, governments can accelerate the transition to more sustainable waste management and energy production methods, aligning economic growth with environmental preservation.

The role of public participation and stakeholder engagement in the regulatory process cannot be overstated. Involving communities, industry representatives, and environmental organizations in the development and implementation of regulations ensures that diverse perspectives are considered. This collaborative approach can lead to more effective and equitable policies that address the needs and concerns of all stakeholders, fostering a sense of shared responsibility for environmental stewardship.

Education and awareness-raising initiatives are also integral to promoting environmental considerations and regulatory compliance. By informing the public and industry about the importance of sustainable practices and the regulations that govern them, these initiatives can drive collective action and support for environmental goals. Workshops, seminars, and informational campaigns can demystify complex regulatory requirements and highlight best practices, empowering individuals and organizations to contribute to a sustainable future.

As the field of waste management and energy production continues to evolve, so too must the regulatory frameworks that govern it. Continuous research and innovation are essential for developing new technologies and practices that enhance environmental performance. Policymakers must remain adaptable, updating regulations to reflect scientific advancements and emerging environmental challenges. This dynamic approach ensures that regulations remain relevant and effective in protecting the environment while supporting progress.

International collaboration and knowledge sharing are crucial for strengthening regulatory frameworks globally. By learning from the successes and challenges of other countries, policymakers can develop more effective regulations that address local conditions while aligning with international standards. Collaborative efforts can also drive innovation and capacity-building, enabling countries to adopt and implement best practices in environmental governance.

Ultimately, the integration of environmental considerations and regulatory frameworks into waste management and energy production is essential for achieving sustainable development. By prioritizing the protection of ecosystems and public health, while fostering innovation and economic growth, these frameworks provide a roadmap for a cleaner, more resilient future. As we navigate the complex interplay between industrial progress and environmental stewardship, the commitment to sustainable practices and robust regulations will be key to ensuring a harmonious coexistence between humanity and the natural world.

Case Studies: Effective Waste-to-Energy Implementations

Across the globe, communities are turning to waste-to-energy (WTE) technologies as a viable solution for managing waste and generating energy. These case studies highlight the diversity and success of WTE implementations, providing valuable insights into their operation and impact. By examining these examples, beginners can gain a better understanding of how WTE projects are executed and the benefits they can offer.

The city of Copenhagen, Denmark, stands as a beacon of WTE success with its Amager Bakke facility, also known as Copenhill. This state-of-the-art plant combines waste incineration with energy recovery, producing enough electricity and district heating to serve tens of thousands of homes. What sets Copenhill apart is its commitment to sustainability and innovation. The facility is equipped with cutting-edge filtration systems that drastically reduce emissions, ensuring compliance with stringent European environmental standards. Moreover, Copenhill's design includes a recreational rooftop complete with a ski slope and hiking trails, making it a unique urban attraction. This integration of functionality and recreation demonstrates how WTE projects can enhance community life while addressing waste management challenges.

Another successful implementation is found in Singapore, where the Semakau Landfill represents a pioneering approach to sustainable waste management. While not a traditional WTE facility, Semakau Landfill incorporates waste-to-energy principles by reducing the volume of municipal waste through incineration before disposal. The ash produced is then transported to the landfill, which is designed to minimize environmental impact. Semakau Landfill is an engineering marvel, constructed on reclaimed land and featuring a rich ecosystem of mangroves and coral reefs. This project highlights the potential for integrating waste management with environmental conservation, creating a harmonious balance between urban development and nature.

In Japan, the city of Kobe has implemented a successful gasification project, converting municipal waste into syngas for energy production. The Kobe facility utilizes a high-temperature gasification process that significantly reduces emissions compared to traditional incineration. The syngas produced is used to generate electricity, contributing to the city's energy supply. Kobe's approach demonstrates the versatility of gasification technology in managing urban waste, offering a cleaner alternative to incineration while extracting valuable energy from waste materials.

The United States also offers noteworthy examples of WTE success, particularly in the county of Montgomery, Maryland. The Montgomery County Resource Recovery Facility employs mass-burn technology to convert waste into energy, providing electricity for thousands of homes and businesses. This facility exemplifies the economic and environmental benefits of WTE technology by significantly reducing landfill dependency and greenhouse gas emissions. The project also includes a robust recycling program, ensuring that recoverable materials are diverted from the waste stream before incineration. Montgomery County's comprehensive waste management strategy illustrates the importance of integrating WTE with other sustainable practices to maximize resource recovery and minimize environmental impact.

In the Indian city of Pune, a decentralized approach to WTE has been implemented through the use of biogas plants. These small-scale facilities process organic waste from households and markets, producing biogas for cooking and electricity generation. Pune's model demonstrates the potential of biogas technology in urban areas, providing a sustainable solution for organic waste management while empowering local communities. By leveraging public-private partnerships and community involvement, Pune's decentralized WTE system is a testament to the power of collaboration in achieving sustainable development goals.

The city of Stockholm, Sweden, offers another compelling case study with its extensive district heating network powered by WTE plants. Stockholm's system integrates multiple WTE facilities, converting municipal waste into heat for the city's district heating network. This approach not only provides a reliable and efficient energy source but also supports Sweden's commitment to reducing carbon emissions and promoting renewable energy. Stockholm's experience underscores the importance of infrastructure and policy support in successfully implementing WTE projects at scale.

These case studies demonstrate the diverse applications and benefits of WTE technologies across different contexts and scales. From large-scale urban facilities to decentralized community projects, WTE implementations can be tailored to meet the specific needs and resources of a given area. The key to success lies in careful planning, stakeholder engagement, and a commitment to sustainability and innovation.

For beginners looking to explore WTE opportunities, these examples offer valuable lessons and inspiration. Understanding the local context, including waste composition, energy demand, and regulatory environment, is crucial for designing effective WTE solutions. Engaging stakeholders, from government agencies to local communities, ensures that projects align with broader sustainability goals and address the concerns and needs of all parties involved.

Moreover, the integration of WTE with other waste management practices, such as recycling and composting, enhances the overall efficiency and impact of waste management systems. By prioritizing resource recovery and minimizing waste, communities can move closer to a circular economy where waste is seen as a valuable resource rather than a burden.

The adoption of advanced technologies and best practices, as demonstrated by these case studies, is essential for optimizing

WTE processes and reducing environmental impact. Continuous research and innovation are needed to improve efficiency, reduce costs, and expand the range of materials that can be processed through WTE technologies. By staying informed about the latest developments and learning from successful implementations, beginners can contribute to the advancement and success of WTE projects.

In conclusion, the effective implementation of waste-to-energy technologies offers numerous benefits, from reducing landfill dependency and greenhouse gas emissions to providing sustainable energy and enhancing community well-being. By examining successful case studies, beginners can gain insights into the diverse applications and strategies that make WTE projects successful. As we continue to explore and develop these technologies, they will play an increasingly vital role in building a sustainable and resilient future. Through continued collaboration, innovation, and commitment to sustainability, WTE can become a cornerstone of modern waste management and energy production systems.

Chapter 5: Circular Economy and Sustainable Resource Management

Principles of the Circular Economy

A circular economy represents a transformative approach to production and consumption, breaking away from the traditional linear model of "take, make, dispose." It aims to redefine growth by focusing on positive society-wide benefits, emphasizing the need for a regenerative system that minimizes waste and makes the most of resources. By delving into the principles of the circular economy, we can uncover strategies to create sustainable systems that benefit both the environment and the economy.

At the heart of the circular economy is the concept of designing out waste and pollution. This principle challenges the conventional notion that waste is an inevitable byproduct of economic activity. Instead, it emphasizes the importance of considering waste as a design flaw. By reevaluating how products are created, companies can integrate strategies that eliminate waste from the onset. This can involve selecting materials that are renewable, durable, and recyclable, as well as designing products for longevity, repairability, and eventual disassembly. The goal is to create systems where everything is used optimally and waste is minimized, if not entirely eliminated.

The second principle revolves around keeping products and materials in use. This involves not only recycling materials at the end of a product's life but also extending the life of products through reuse, remanufacturing, and refurbishment. For example, the electronics industry is increasingly adopting modular designs that allow components to be replaced or upgraded easily, thereby extending the life of devices and reducing the need for new resources. Similarly, the fashion industry is exploring innovative business models like clothing rental and resale platforms to keep garments in circulation for longer. By keeping materials in use, the circular economy reduces the demand for new resources and lessens environmental impact.

Regenerating natural systems is the third principle, emphasizing the need to restore and enhance ecosystems rather than depleting them. Unlike the linear economy, which often results in resource extraction and environmental degradation, the circular economy seeks to build natural capital. This can involve practices such as regenerative agriculture, which enhances soil health and biodiversity, or urban planning strategies that incorporate green spaces and nature-based solutions. By adopting regenerative practices, we can ensure that natural systems are preserved and enhanced for future generations, contributing to the overall resilience and sustainability of our planet.

The transition to a circular economy also requires a shift in consumer mindset and behavior. Individuals play a crucial role in this transformation, as their choices drive demand for sustainable products and services. Encouraging consumers to prioritize quality, durability, and sustainability over low cost and disposability is essential. This shift is supported by increasing awareness and education about the environmental impact of consumption and the benefits of circular practices. Businesses and governments can also incentivize sustainable choices through policies, labeling schemes, and financial incentives, creating an environment where circular options are accessible and attractive to consumers.

Collaboration across sectors and stakeholders is vital for the successful implementation of a circular economy. No single entity can achieve circularity alone; it requires coordinated efforts among governments, businesses, academia, and civil society. Collaborative platforms and partnerships can facilitate knowledge sharing, innovation, and the development of circular solutions that address common challenges. For example, industries can work together to establish shared infrastructure for resource recovery and recycling, while research institutions can provide insights into new materials and technologies that support circular practices. By fostering a culture of collaboration, we can accelerate the transition to a circular economy and unlock its full potential.

Technology and innovation are key enablers of the circular economy, providing the tools and solutions needed to implement circular practices effectively. Advances in digital technology, such as the Internet of Things (IoT) and blockchain, can enhance transparency and traceability in supply chains, enabling businesses to track the lifecycle of products and materials. This information can inform decisions about reuse, recycling, and waste management, ensuring that resources are used efficiently. Additionally, innovations in materials science, such as biodegradable plastics and advanced recycling techniques, offer new opportunities to reduce waste and enhance the sustainability of products.

The economic benefits of a circular economy are significant, offering new business opportunities and driving growth while reducing environmental impact. By decoupling economic activity from resource consumption, businesses can achieve greater resilience and competitiveness. Circular practices can also lead to cost savings through more efficient resource use and waste reduction. For example, businesses that adopt circular models may experience lower material costs, reduced waste disposal fees, and increased revenue from new services and product lines. On a macroeconomic level, the circular economy can stimulate job creation and innovation, contributing to a more sustainable and inclusive economy.

However, transitioning to a circular economy is not without challenges. It requires significant changes to existing systems and infrastructure, as well as overcoming regulatory and financial barriers. Policymakers have a crucial role in facilitating this transition by creating supportive regulatory frameworks and providing incentives for circular practices. This can include measures such as extended producer responsibility, which holds manufacturers accountable for the entire lifecycle of their products, or tax incentives for businesses that invest in circular solutions. By creating an enabling policy environment, governments can drive the adoption of circular practices and support the development of a circular economy.

Education and capacity-building are also essential for empowering individuals and organizations to embrace circular practices. By providing training and resources, we can equip people with the skills and knowledge needed to implement circular solutions in their communities and industries. This includes fostering a mindset of innovation and continuous improvement, where individuals are encouraged to experiment with new ideas and approaches. By investing in education and capacity-building, we can create a culture of sustainability and innovation that supports the transition to a circular economy.

In conclusion, the principles of the circular economy offer a roadmap for creating a sustainable future where resources are used efficiently, waste is minimized, and natural systems are restored. By embracing these principles, businesses, governments, and individuals can contribute to a more resilient and equitable society. The journey towards a circular economy requires collaboration, innovation, and a commitment to change, but the rewards are well worth the effort. As we continue to explore and develop circular solutions, we can build a world where economic growth and environmental sustainability go hand in hand.

Designing Products for a Circular Life Cycle

Designing products for a circular life cycle requires rethinking the traditional approach to product development, focusing on sustainability and resource efficiency at every stage. This paradigm shift not only addresses the growing environmental impact of consumer goods but also unlocks new economic opportunities by extending product longevity and reducing waste. By incorporating circular principles into design processes, companies can create products that contribute to a sustainable future while meeting consumer demands.

The first step in designing for a circular life cycle involves selecting materials that align with circular principles. Materials should be chosen based on their renewability, durability, and recyclability. Renewable materials, such as bioplastics and sustainably sourced wood, reduce dependence on finite resources and promote environmental sustainability. Durability ensures that products withstand wear and tear, extending their useful life. Recyclability allows materials to be reclaimed and reused at the end of a product's life, minimizing waste and preserving resources. By prioritizing these attributes, designers lay the foundation for products that support a circular economy.

Modular design is another key aspect of circular product development. By creating products with interchangeable components, companies enable easier repair, upgrading, and customization. This approach not only extends the product's lifespan but also enhances user experience by allowing consumers to adapt products to their evolving needs. For example, modular smartphones allow users to replace or upgrade individual parts, such as batteries or cameras, rather than discarding the entire device. This flexibility reduces electronic waste and empowers consumers to make sustainable choices.

Incorporating closed-loop systems into product design can further enhance circularity. Closed-loop systems consider the entire life cycle of a product, from raw material extraction to end-of-life disposal, ensuring that all materials are either reused or recycled. Designers can achieve this by integrating take-back programs, where companies collect used products for refurbishment or recycling. These programs not only create a continuous flow of materials but also strengthen customer relationships by demonstrating a commitment to sustainability. By designing products with closed-loop systems in mind, companies can reduce their environmental footprint and contribute to a more sustainable economy.

The role of digital technology in circular product design cannot be overstated. Digital tools, such as 3D modeling and simulation, enable designers to experiment with different materials and configurations, optimizing products for circularity before they reach production. Additionally, digital platforms can facilitate the tracking of products throughout their life cycle, providing valuable data on usage patterns and material recovery. This information can inform future design decisions and improve the efficiency of recycling processes, supporting the transition to a circular economy.

Consumer engagement is crucial for the success of circular products. Educating consumers about the benefits and functionality of circular products can drive demand and adoption. Companies can leverage marketing campaigns, product labeling, and educational initiatives to communicate the advantages of circular products, such as cost savings, environmental impact, and product longevity. By fostering a deeper understanding of circular principles, companies can empower consumers to make informed purchasing decisions and support the growth of a circular market.

Partnerships and collaboration play a vital role in advancing circular product design. By working with suppliers, manufacturers, and recycling facilities, companies can create integrated supply chains that support circular practices. Collaborative efforts can lead to the development of new materials, processes, and business models that enhance product circularity. For instance, partnerships between electronics manufacturers and recycling companies can streamline the recovery and reuse of valuable materials, closing the loop on electronic waste. By fostering a culture of collaboration, companies can accelerate the adoption of circular design principles and drive systemic change.

Designing for a circular life cycle also requires a shift in business models. Traditional linear models, which rely on high sales volumes and planned obsolescence, are incompatible with circular principles. Instead, companies must explore alternative models that prioritize value over volume. Subscription services, product-as-a-service, and leasing arrangements are examples of business models that support circularity by focusing on product performance and longevity rather than ownership. These models create new revenue streams and build stronger customer relationships while reducing the environmental impact of production and consumption.

Policy and regulation can provide the framework and incentives needed to support circular product design. Governments can implement policies that promote circular practices, such as extended producer responsibility, which holds manufacturers accountable for the entire life cycle of their products. Tax incentives and subsidies can encourage companies to invest in circular innovations and infrastructure. Additionally, standards and certifications can help consumers identify and choose circular products, further driving demand. By creating an enabling policy environment, governments can support the transition to a circular economy and encourage widespread adoption of circular design principles.

The integration of circular design principles into product development offers numerous benefits, from reducing environmental impact and conserving resources to enhancing brand reputation and customer loyalty. By reimagining products as part of a continuous loop rather than a linear trajectory, companies can create sustainable, innovative solutions that meet the needs of both consumers and the planet. As more companies embrace circular design, the cumulative impact will contribute to a more resilient, sustainable economy that benefits all stakeholders.

In conclusion, designing products for a circular life cycle represents a fundamental shift in how we create and consume goods. By prioritizing sustainability, durability, and resource efficiency, companies can create products that support a circular economy and address pressing environmental challenges. Through collaboration, innovation, and consumer engagement, the principles of circular design can drive meaningful change and pave the way for a sustainable future. As we continue to develop and refine these concepts, the potential for a circular economy grows, offering a pathway to a more sustainable and equitable world.

The Role of Extended Producer Responsibility

Extended Producer Responsibility (EPR) has emerged as a pivotal policy approach in the quest for sustainable waste management and resource conservation. By holding producers accountable for the entire lifecycle of their products, EPR encourages companies to design environmentally friendly products and manage their end-of-life waste. This chapter examines the multifaceted role of EPR in advancing a more sustainable economy, offering insights into its implementation, benefits, and challenges.

The concept of EPR shifts the responsibility of waste management from consumers and local governments back to producers. Traditionally, the burden of managing product waste has fallen on municipalities, leading to inefficiencies and increased costs. EPR, by contrast, incentivizes producers to take a proactive role in minimizing the environmental impact of their products. Companies are encouraged to design products that are easier to recycle, use fewer resources, and contain less hazardous materials. This shift in responsibility fosters innovation and drives the development of sustainable products.

One of the most significant benefits of EPR is its potential to reduce waste and promote recycling. By requiring producers to manage the end-of-life disposal of their products, EPR creates a strong incentive to design for recyclability. Products that can be easily disassembled and recycled are more likely to be recovered and repurposed, reducing the amount of waste sent to landfills. In turn, this conserves resources and reduces pollution, contributing to a more sustainable environment. The increased demand for recyclable materials also stimulates the growth of recycling industries, creating jobs and economic opportunities.

EPR programs can take various forms, depending on the specific goals and regulatory frameworks of each region. Some EPR initiatives require producers to establish and fund their own recycling programs, while others set recycling targets that companies must meet. In some cases, producers may form collective organizations, known as Producer Responsibility Organizations (PROs), to pool resources and manage waste collectively. These organizations can achieve economies of scale, making recycling and waste management more efficient and cost-effective.

The electronics industry offers a compelling example of EPR in action. With the rapid pace of technological advancement, electronic waste, or e-waste, has become a significant environmental challenge. EPR programs targeting e-waste have been implemented in many countries, requiring producers to establish take-back schemes for old devices. These programs not only facilitate the recovery of valuable materials, such as gold, silver, and copper, but also ensure the safe disposal of hazardous substances like lead and mercury. By holding producers accountable for e-waste, EPR has spurred innovation in product design and recycling technologies, reducing the environmental impact of electronic products.

Packaging waste is another area where EPR has made significant strides. Packaging materials, such as plastics, paper, and glass, constitute a substantial portion of municipal solid waste. EPR programs for packaging require producers to contribute to the cost of recycling and disposal, incentivizing them to reduce packaging waste and improve recyclability. As a result, many companies have adopted sustainable packaging solutions, such as lightweight materials, biodegradable options, and reusable packaging systems. These innovations not only reduce waste but also appeal to environmentally conscious consumers, enhancing brand reputation and competitiveness.

Despite its many benefits, implementing EPR programs can present challenges. Establishing effective regulatory frameworks and ensuring compliance requires coordination among various stakeholders, including government agencies, producers, and consumers. Additionally, the financial burden of EPR may be passed on to consumers through higher product prices, potentially affecting affordability and accessibility. To address these challenges, policymakers must carefully design EPR programs that balance environmental goals with economic considerations, ensuring that they are fair, transparent, and effective.

Consumer awareness and participation are crucial for the success of EPR initiatives. Educating consumers about the importance of recycling and the role of EPR can drive demand for sustainable products and increase participation in take-back schemes. Clear labeling and communication can help consumers make informed choices and properly dispose of products at the end of their life. By fostering a culture of responsibility and sustainability, EPR programs can empower consumers to play an active role in waste management and resource conservation.

International collaboration can also enhance the effectiveness of EPR programs. As supply chains become increasingly global, harmonizing EPR regulations across borders can reduce complexity and facilitate compliance for multinational companies. Sharing best practices and experiences can help countries design and implement effective EPR programs, accelerating the transition to a circular economy. Collaborative efforts can also address the challenges of e-waste and packaging waste, which often cross national boundaries.

In addition to regulatory measures, voluntary EPR initiatives have gained traction as companies recognize the value of sustainability. Many businesses are taking proactive steps to reduce their environmental impact by adopting EPR principles, even in the absence of formal regulations. These voluntary efforts not only demonstrate corporate responsibility but also provide a competitive advantage by appealing to environmentally conscious consumers and investors. By leading the way in sustainability, companies can influence industry standards and drive positive change.

The role of EPR in promoting a circular economy cannot be overstated. By extending responsibility for product waste to producers, EPR encourages the development of sustainable products and systems that keep materials in use for as long as possible. This shift from a linear to a circular economy reduces reliance on finite resources, minimizes waste, and mitigates environmental impact. As more countries and industries embrace EPR, the potential for a sustainable and resilient economy grows.

In conclusion, extended producer responsibility plays a critical role in advancing sustainable waste management and resource conservation. By holding producers accountable for the lifecycle of their products, EPR incentivizes innovation and the development of environmentally friendly solutions. Through collaboration, education, and effective regulation, EPR can drive the transition to a circular economy, creating a more sustainable future for all. As we continue to refine and expand EPR initiatives, they will remain a cornerstone of modern environmental policy, shaping the way we produce, consume, and manage resources.